THE FACE ON
MARS

THE FACE ON MARS

Randolfo Rafael Pozos

Chicago Review Press • **814 North Franklin** • **Chicago**

For Teachers
Everywhere
Who Tend the Sacred Flame

Especially for
S. Christa McAuliffe
of the Challenger Seven

Who closed her classes over the years
with a message for the ages:

"To the Stars."

CONTENTS

FOREWORD

This work by Dr. Pozos is the first report of a remarkable story that began in 1979. It was then that two researchers, Vincent DiPietro and Greg Molenaar, discovered an unusual feature on two images of the surface of Mars that had been taken by one of our Viking orbiter spacecraft, sent to the planet in 1976 for the primary purpose of discovering whether or not life existed there. These images showed what appeared to be a large carving of a humanoid face lying on the ground and staring into space.

The two researchers published the pictures and were surprised by the emotional reaction that resulted. This came not from the general public, but from the planetary science community: the very people they had hoped would bring their expertise to bear on the enormous issues raised by the possibility that an ancient civilization might once have lived on Mars. To their shock they instead found themselves under attack for daring to raise the question. The images were simply disposed of with the comment that they were nothing more than an illusion brought about by light and shadow. With their integrity now in question, Molenaar and DiPietro set to work to show that this was not so. They even devised a new method of increasing the resolution of the digital images (taken from high above the planet) so that the eye sockets were plainly visible and by means of false color and computer enhancement, what appeared to be an eyeball could be seen in the right socket. All to no avail. The two researchers found themselves shunned by the planetary scientific community. Finally, after four years and thousands of dollars of their own money, tired of the constant bickering, they decided to lie low.

In 1983, science writer and analyst Richard Hoagland saw their work and started reviewing additional images of the area and its surrounds. He became more and more convinced that as well as the "Face", there were other unusual surface features on the planet and that a full scale scientific investigation of these abnormalities was called for. He, like DiPietro and Molenaar, quickly found that those scientists that should have been the most interested were instead the most reluctant to accept that there was anything on the images other than natural features. Some, however, did admit that there were interesting anomalies and a few even said they would like to take part in further research on the question, but after a week or two, all of them withdrew. Hoagland and some supporters, then decided to have the features examined by some other jury. He figured that if the planetary scientists would not cooperate, surely other scientists in other disciplines would.

Dr. Pozos, an anthropologist, was one of the first to be approached. He has prepared this initial account of the early experiences and conclusions of the team that was set up to review the images, and has included some thoughts of his own regarding the significance of their findings. It makes fascinating reading, both as an account of the investigation itself, and as a study of the interaction between a group of scientifically oriented people (most of whom had never met) conducting an ongoing dialogue on a highly speculative issue by means of a computer conference.

One does not have to agree with all the findings to feel both an involvement with the process itself, as well as with the subject matter. After seeing the images and reading the report, it is difficult to believe that all the unusual features that are outlined can be natural. Sufficient reasonable doubt appears to exist to require an early return mission to the planet to determine the truth. Is this likely to happen? Not if one judges by the continuing reaction from some members of the planetary community. Instead of treating the issue objectively as befits a scientist, they insist on making personal attacks on the character and integrity of some of the team members. A strange way to act.

Dr. Pozos has performed a great favor in publishing this report and letting us judge for ourselves about the need for further investigation of these unusual features. He, Richard Hoagland, and the entire team deserve our support and encouragement in their continuing effort to have these images properly reviewed. Discovery of an ancient civilization on Mars would raise issues of enormous importance to our own civilization, not the least of which would be the question of why the civilization died out, and why and how Mars lost its water and atmosphere.

In view of the importance of such a question to our own future here on Earth, it appears all the more strange that the team should have had such difficulty in getting support from the planetary community for their guest. Especially when this same community is spending millions of dollars and considerable effort in their radio search of nearby galaxies for signs of extra-terrestrial intelligence. Should the results of the Independent Mars Investigation Group eventually prove that intelligent life in a form similar to our own once existed on Mars, it would most surely indicate that just as war is too important to be left to the generals, so science is too important to be left only to the scientists!

David C. Webb, Ph.D.
Member, President's
Commission on Space

ACKNOWLEDGEMENTS

Many people extended considerable time and effort in the creation and carrying out of this scientific inquiry. The primary supporters of the research deserve special mention.

C. Renwick Breck of the InfoMedia Corporation generously provided the computer conferencing services of Notepad. Lambert Dolphin and Paul Shay of SRI International played a critical role in the organization of the inquiry, in addition to providing extensive photographic and imaging services. The cooperation and participation of Vincent DiPietro and Gregory Molenaar, who initiated the first serious inquiry over ten years ago, were a primary impetus in the research.

I owe a special debt of thanks to the founders of the Mars Project, Richard C. Hoagland, Brian O'Leary (scientist/astronaut), and David Webb (member of the President's Space Commission) for their assistance and encouragement, to Thomas Rautenberg and C. West Churchman of the University of California, Berkeley Mars Investigation Group for their thoughts and ideas.

As in any scientific endeavor, this book and the research it records was made possible by innumerable researchers whose patient efforts are quietly stored away. We owe them an immense debt as a society, since they have made the future possible.

Special thanks are in order to the National Aeronautics and Space Administration (NASA) and the United States Geological Survey (USGS) for their assistance in providing images and data tapes despite the fact that they did not share our research interest.

Despite all of the correction and guidance I received from many scientists, any errors which remain are my own.

Special thanks are well deserved by my forbearing family and friends. My wife and fellow anthropologist Kathleen kept the writing of the book organized and on-track. Richard C. Hoagland and Linda Matthews did their best to clarify some sections and translate others into readable English. C. Renwick Breck allowed me to save several weeks by using Notepad to transfer the manuscript electronically from my computer to Chicago and to typesetting.

To all of these friends and associates, my thanks.

CHAPTER ONE

Adventures on a Notepad

It all started one quiet Sunday afternoon, July 24, 1983. C. Renwick Breck, the vice president of InfoMedia, a computer conferencing company, picked up his phone. When I heard Ren's voice across the line, I knew there was urgency, primarily because for him voice communication over telephone lines is slightly old fashioned and quaint, something reserved for personal conversations or to summon wayward computer conferencers such as myself to an important meeting hosted by InfoMedia's computer near the San Francisco Airport in San Bruno.

Ren and I had met in 1974 in Berkeley through the Futures Planning Council. I was a graduate student in anthropology. Over the next several years we got each other into various volunteer projects using appropriate new technologies to solve problems of urban housing, health, and other human needs.

When I picked up the phone, I had a vague sense of apprehension, perhaps it was a slightly guilty conscience. Several months earlier I had gotten Ren involved with the Nepal Blindness Project. This worthy cause needed a small favor that Ren could provide: a specially equipped transport helicopter to treat and prevent blindness in Nepal, on the roof of the world in the Himalayas. Of course the Nepal story could be a book in itself. It is enough to say that the project had been successful. Now he was returning the "favor."

"I have a friend who needs to talk to an anthropologist immediately." The words came across the slightly antiquated voice connection somewhat like a wayward line out the Book of Common Prayer, a style which Ren reserves for solemn occasions.

As an anthropologist interested in high technology transfer and the future development of human society, I had also developed a career as a health care planner and administrator, a perfect laboratory and observation post for studying the relationship between new technologies and the forces and trends of urban social development.

Someone needing "an anthropologist immediately" struck me as one of Ren's best lines. In my daily work with physicians I had learned what an

Viking I liftoff atop Titan/Centaur rocket, August 20, 1975. The spacecraft was sterilized by NASA to get rid of microooganisms from Earth before its flight to Mars. [NASA]

1

Artist's conception of the Red Planet prior to the Viking program. Since Percival Lowell's death in 1916, intelligent life on Mars was considered possible if not likely. When NASA reluctantly published its conclusions from the Viking program, interest in Mars among the general public declined sharply. [NASA]

emergency really is. That someone ''needed an anthropologist'' was at least clever. The anthropologist who digs up the past — the archaeologist — usually arrives fashionably late by at least several centuries.

Social and cultural anthropologists spend years trying to understand the dynamics of current or past societies by studying the language and behavior of humans and similar primates such as monkeys, chimpanzees, and apes. Even those anthropologists who study the exotic rituals and magic of American corporations take weeks to produce their studies of the human design of these mysterious groups lost in the shadowy forests of skyscrapers.

The only thing that I could think of was that I had really done it to Ren with the Nepal Blindness project. Several thoughts came to mind: another low income housing project?, outreach services for the poor elderly in the downtown hotels?, an independent living project for the handicapped? I was scanning my memory banks when my mind registered an input/output error.

''Richard Hoagland, the former science advisor to Walter Cronkite, thinks that he may have found *the remains of a lost city on Mars.*''

I knew that Ren had been working hard on demanding projects (such as the use of FM carrier radio waves to transmit printed text which people could read at home on their television sets). However, from the tenor of his voice I knew that I had just lost my Sunday afternoon and a lot of free time in the coming months.

I felt the trap of my own curiosity about to spring but it was worth at least one try at escape. My Berkeley training might now save me. I had seen senior professors use it with great success as they fled unwelcome snares and disappeared into the labyrinth of Kroeber Hall. ''Ren, I am not an archaeologist, it's not my field.''

Ren reminded me of my undergraduate training in philosophy, classics, and biology. He said at this stage Dick Hoagland needed to talk to someone with an open mind and enough background to review the basic questions. A terrestrial anthropologist was needed, Ren insisted, because there was a large humanoid ''face'' looking upward close to what appeared to be pyramids. He went on to make the obvious point that there was as yet no field of Martian anthropology.

Martian Anthropology

Martian anthropology? This was one of Ren's best hooks. Could there be such a field? In order to prepare for my comprehensive exams in anthropology I once had to write statements of "field." In my mind I reviewed the requirements. Usually a field is defined as an area of research which involves a geographic place such as North America, and deals with a group or groups of people such as "Scandinavian immigrants" or a topic like "family structure" or "legal systems."

Mars is a geographical place, with a natural history which we are beginning to see in its major outlines. Discovery of artifacts such as a large sculpture or building would indicate that there once may have been a society there. However, even if the face and the other unusual surface features of Mars were not the product of natural processes, could we be sure that the creators of these surface features were members of our own species?

If and when anthropologists begin to study other intelligent and symbol making life forms on other worlds, anthropology will have to change its name. *Anthropos* in Greek refers to humans or *Homo sapiens*. The study of the these advanced life forms might be called "interplanetary sociology."

As I rolled the concept of Martian anthropology over in my mind it occured to me that this question of the face on Mars, or should I say the challenge of the face on Mars, showed very clearly the meeting place of the sciences and the humanities.

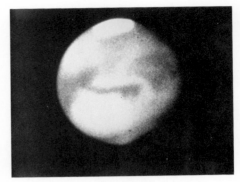

Viking I photo taken on approach to Mars. White polar ice cap appears seasonally, as do giant dust storms which obscure the surface. [NASA]

Viking I photo of the northern hemisphere of Mars. The Vikings' cameras were designed primarily for mapping the planet and locating a good place to land. [NASA]

The sciences and the humanities represent a recent branching of Western European thought, which used to be called "natural philosophy" until about 100 years ago. It used to be thought that we could understand the nature and purpose of the universe and ourselves by a close observation of the behavior of nature and human society.

Along the way, we developed great tools and machines based on our observations, along with effective techniques for guiding and directing human behavior. Cars, telephones, computers, marriage counseling, special education for the handicapped, and human rights are only a few of the conceptual or practical tools which have improved the quality of human existence. This digression into applied technology was so rewarding that much of the original quest was abandoned.

The challenge of the face on Mars rekindles the earlier question by bringing us back to where we started observing the behavior of "nature and human society."

We have a series of light and dark spots on a computer screen or on a piece of film, produced by scanning a neighboring planet. Gradually our minds begin constructing the image of something that looks like a face. Is it a human face? The imaging specialists disagree among themselves. Some say that it is more than just a trick of light and shadow. Others say "no" and claim that like many images on earth it is only the product of the right lighting and an active human imagination.

Some scientists who study the image are distracted by speculation. Others won't permit themselves to speculate. Some are fearful of becoming scientific heretics and losing their jobs and their identities. Others feel that science shouldn't or can't deal with basic philosophical or ultimate questions such as: What is it to be human? Why are we here? Where are we going?

High-altitude photograph of Cydonia in the northern hemisphere of Mars taken by the Viking I orbiter in 1976. At the upper right is the unusual surface feature known as the Face, measuring approximately one mile in diameter. At the upper left is a five-sided pyramidal structure which is approximately three times larger than the Great Pyramid in Egypt. [NASA]

The Riddle of the Martian Sphinx

The challenge of the face on Mars is not unlike the riddle of the ancient Sphinx. "What is it that crawls on four legs in the morning, walks on two at midday, and walks on three in the evening?" According to the ancient myths people who tried to answer the riddle and failed were killed by the Sphinx. To answer the riddle correctly a basic knowledge of our species is necessary. Humans crawl on all fours before they learn to walk on two legs and often need the support of a cane or third leg in old age.

The ancient hero answers the riddle of the Sphinx with one word, *anthropos*, humanity. Until recently we always used to translate the answer as "man," even though anthropos is the inclusive term. We have attempted to understand ourselves by calling ourselves man and by denying or avoiding that dimension of the human called woman.

In many ways the face on Mars is a present day Sphinx. Hundreds of thousands of tiny dots or picture elements called pixels have given us something that seems to correspond to our image of ourselves. The only problem is that this face is on our neighboring planet. Our exploration of our solar system and outer space is a colossal extension of our sometimes fatal but always necessary human curiosity. What are we looking for in all of this?

The anthropology of Mars is, at this point, the study of the attempt by humans to deal with the challenge of the face on Mars. Is the face there? If so what does it mean? If not, what does this mistaken human perception tell us about human perceptions and the concept of being human in the late twentieth century?

In a broader sense, we can say that the anthropology of planetary and deep space exploration is the "attempt to interpret the meaning and the context of the human riddle, as humans chart the universe trying to understand themselves."

As I sat in my chair dialing Richard Hoagland's phone number, I had a curious feeling that I might be returning to the armchair anthropology of the nineteenth century, taking down accounts of far off lands from missionaries and journalists. Little did I know or even guess that the information would come from satellites or that we would examine the accounts in the electronic drawing room of a computer, through microcomputers in our homes and offices. Our armchairs would become less comfortable as we wrestled with the challenge of the face.

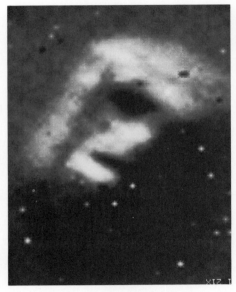

Closeup view of the Face on Mars. Scientists are presently studying other similar Faces found in the more than 60,000 images sent back to Earth by the Viking probes. [SRI]

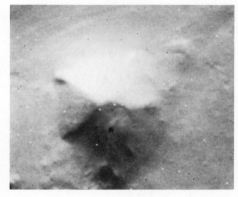

Closeup of one of the Pyramids near the Face. This five-sided pyramid appears to be damaged on two sides. [SRI]

Napoleon's soldiers exploring the Sphinx and Pyramids of Egypt. [Engraving from Description de l'Egypt, *1825; Thomas Fisher Rare Book Library, University of Toronto]*

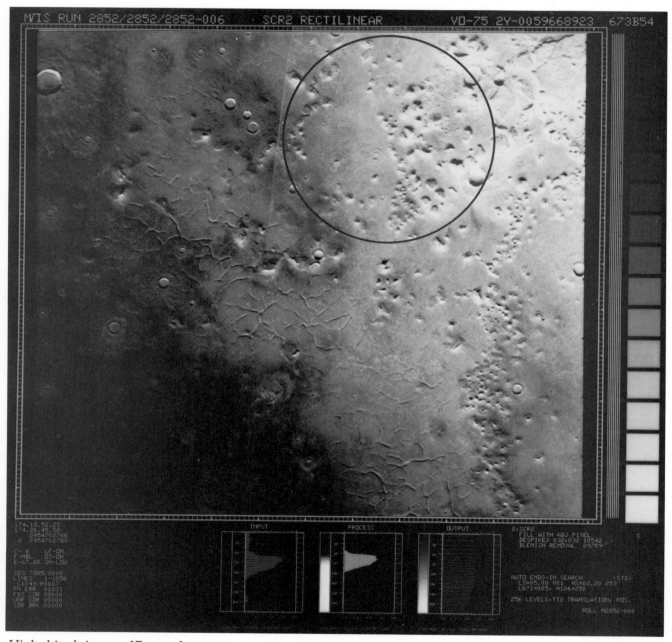

High altitude image of Face and Pyramids [upper left]. NASA 673B54.

CHAPTER TWO

Science and Extraterrestrial Intelligence

It was a short drive from my home in Alameda to Richard Hoagland's cottage on the border between Oakland and Berkeley. The Elmwood district was familiar to me from my student days. The modest bungalows and the imposing brown shingled houses are home to many academics, researchers, writers and artists.

My wife, Kathleen, and I had lived only a few blocks away when we were graduate students. Kathleen is a medical anthropologist and has done research on the behavioral and cultural components of chronic pain in American women. It was with a certain sense of nostalgia that we returned to the streets and avenues we had walked so many evenings. Martian anthropology had never crossed my mind until a few minutes before, and now I was trying to summon one of those evening discussions Kathleen and I had had on these very streets, about the future dimensions of human society as it adapted to planetary exploration.

Richard Hoagland had been doing research on the debris that swarms around the rings of Saturn, as a follow-up to his hypothesis about the presence of a remarkable high energy ''thing in the ring'' discovered by Dr. James Warwick during the Voyager unmanned mission to the planet in 1980.

Hoagland was looking for techniques which could clarify or enhance electronic images from planetary probes such as Mariner and Viking. A better look at the tiny moons of Saturn might provide some additional clues. He had sent away for a monograph published by two imaging specialists, Vincent DiPietro and Gregory Molenaar. DiPietro and Molenaar had been with the Goddard Space Flight Center in Maryland and had developed their own copyrighted method for enhancing images called the ''Starburst Pixel Interleaving Technique,'' or SPIT process.

DiPietro and Molenaar in 1979 had applied this technique to the ''Face'' on Mars, using two images taken on different days from different angles, and had come to the conclusion that the ''Face'' is an actual surface feature and not just a creation of light and shadow. (Hoagland had been at the Jet Propulsion Lab in Pasadena as a journalist the day the image of the face had come across the screen and like everyone else he had accepted the

Monograph by Vincent DiPietro and Gregory Molenaar, co-developers of the SPIT process. This monograph can be purchased from Mars Research, P.O. Box 284, Glenn Dale MD 20769.

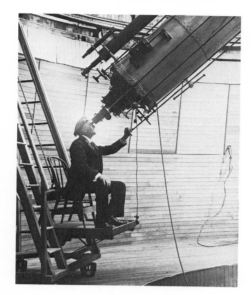

Percival Lowell looking through 24-inch refractor telescope. [Lowell Observatory]

explanation that when the area had been passed over again the face was not there.)

DiPietro and Molenaar presented their findings in 1980 at professional meetings and circulated their monograph. The reception was sometimes polite and sometimes hostile but the idea never really caught on. This is not unusual in science. New ideas can be ignored for years or even generations.

Resistance among scientists to evidence of life on Mars has an important cultural and historic component.

Much of the great twentieth century American momentum and interest in planetary exploration derived from the fact that Percival Lowell, the brother of poet Amy Lowell, managed to excite European and American popular interest in life on Mars. This is a long story with which we will deal later. However, the main story line goes like this. Using the best technology of his time, a large visual telescope, Lowell became convinced that the long rod-like markings on Mars which expand and contract with the Martian polar ice caps represented artificial waterways highlighted by vegetation waxing and waning in seasonal variations on a planet with conditions similar to earth. The changing colors of the planet during the seasons — reddish to greenish and black — appeared to indicate chlorophyll-based green plant agriculture. Consequently, the spider web-like markings (called *cannali* or "channels" by their Italian discoverer Schiaparelli) became translated into English as "canals". Although a correct literal translation , the connotation to the English speaker is far different from mere channels!

As a respected and talented patron of science, Lowell's name lended plausibility, in the mind of the public, at least, to H.G. Wells' fictional elaborations. The Martian theme in popular literature became a staple of entertainment in early and mid-twentieth century movie houses. Flash Gordon and other heroes used Mars and other planets as a backdrop for adventures. Classical mythological notions and legends about Mars were blended with twentieth century notions of scientific progress in art deco morality plays.

The success of archaeology and cultural anthropology in the late nineteenth and early twentieth centuries, in making the public aware of other cultures in different times and places (King Tut's tomb discovery and

H.G. Wells, 1866-1946. [Metropolitan Toronto Library Board]

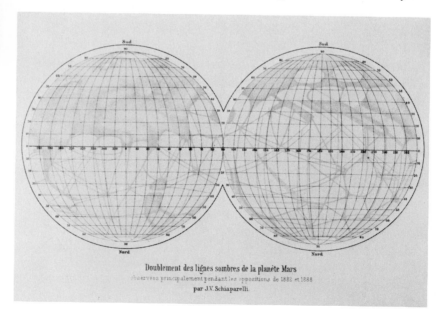

Doublement des lignes sombres de la planète Mars
observées principalement pendant les oppositions de 1882 et 1888
par J.V. Schiaparelli.

Schiaparelli's web-like cannali.

a wave of ''Egypt mania'' in the 1920's) provided exotic costumes, temples, and villains on other planets for the fiction. Ray Bradbury's enduring 1950's classic, *The Martian Chronicles*, would lend its name to our eventual computer conference, which was to become an armchair sequel with its own unexpected twists and turns.)

The upshot of all this was to create a certain public expectation that other worlds and other civilizations were not terribly distant. In terms of space, they might exist close at hand. In terms of time, perhaps only a few human generations would separate us from some type of direct encounter.

Of course, this leads us to remember a whole host of more recent movies after World War II which dealt with invasions from other planets. The more benign possibilities of extraterrestrial encounter came to the fore in the sixties with productions like the television situation comedy *My Favorite Martian*. Certainly, we cannot discount the American cultural phenomenon of Superman as a much earlier example of a beneficial experience with an extraterrestrial. However, Superman's close resemblance to the physical and cultural ideal of the all American hero causes him to be more of a perfected earthling albeit with a weakness for kryptonite and a certain woman reporter.

The television series *Star Trek* represents one of the most sophisticated American cultural images of the future. Here a wide variety of intelligent life forms throughout the cosmos form an intergalactic world which is divided between a totalitarian empire and a federation of free world republics.

The similarities between *Star Trek* and the Cold War of the fifties and sixties are readily apparent. However, *Star Trek* introduces a new element in the fight against cosmic Communism. The heroes of *Star Trek* are ethnically and culturally diverse. Minority men and women play positive technical and support roles either as humans or as other humanoid species. Of course Captain Kirk is the athletic white American male, but overall *Star Trek* exemplifies the desires and fears of a more culturally diversified American society.

The optimistic tone of *Star Trek* contrasts vividly with the determinism of Arthur Clarke's novel and Stanley Kubrick's monumental movie *2001*. Here, extraterrestrial contact is not a future experience for humanity. In essence, human development, from the stone age to the space age and beyond, is the product of yet another intelligence which is not as totally altruistic or transcendent as the God of the Judaeo-Christian tradition. The meaning, purpose, and direction of human physical and cultural development is directed from outside.

In *2001* the wonder of human technical and intellectual achievement, and our apparent lack of control over this development, and its harmful or beneficial applications is exposed as the program of a much more advanced intelligence. All that is and was and will be was meant to be. Given the popularity of Eastern religions and philosophies in the West during the twentieth century, along with the technological wonders and horrors of the century, this is not a surprising view.

A more balanced or perhaps existential view is presented in the movie *E.T.* The archetype of American adventure literature, the young boy approaching puberty is recast. Instead of Huckleberry Finn adrift on a raft on the rivers of the Midwest, a young boy adrift with his single parent family in a newly built California suburb encounters another young boy adrift on Earth.

From these cursory notes on the cultural iceberg of extraterrestrial intelligence in American popular culture two things become evident. First, the solar system and deep space have become new and exotic locales for the age-old questions of western civilization as experienced in the progress and violence of the twentieth century. Second, humans seem unable to fictionally invent a non-earth civilization which is not fundamentally human. Some of the body parts might be different but the loves, the hates, the jealousies have not changed since the stone age.

Paradoxically, the popular acceptance of civilizations on other worlds and unidentified flying objects (UFO's) has increased skepticism in the scientific community about intelligent life forms anywhere in our immediate neighborhood, the Milky Way. Carl Sagan, who has pioneered much of the scientific research on biological life forms on Mars, summarized the current scientific consensus in his bestseller, *Cosmos*. In a few words, the good news is that on the basis of scientific probability there probably are other worlds with intelligent life forms. The bad news is that they are either hundreds of years away (even travelling at the speed of light — 186,000 miles per second — which Einstein claimed to be impossible,) or so much older (implying that they are much farther advanced than we are) that meaningful communication could be impossible.

If there might be other life forms more highly developed than we are why haven't they contacted us? Or at least paid us some type of beneficial visit?

The Search for Extraterrestrial Intelligence (SETI) is a fascinating field of scientific research and speculation attempting to deal with these difficult questions. However, the scientists who are willing to deal with these questions have to tread on very thin scientific ice.

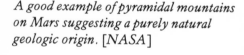

A good example of pyramidal mountains on Mars suggesting a purely natural geologic origin. [NASA]

The major problem: the question of consciousness

Our present notions of correct science grew out of the separation of mind and body in the seventeenth and eighteenth centuries. In earlier centuries and in today's religions, knowledge, intelligence, consciousness, even wisdom are spiritual realities which reside in the human soul or psyche.

The development of "the scientific method" by Descartes started a trend which placed mind and consciousness not in the soul or psyche but in that computer which we call the brain. For Descartes, the human body was a machine operated by a soul. Gradually, for contemporary science this principal of vitalism was discarded, and the human body became a computerized machine.

Consequently, when scientists today attempt to deal with possible extraterrestrial intelligence, their scientific methodology imposes certain real limits because it was not developed to quantify or document non-physical experiences.

The human species has a very long history of dealing with the non-physical or the metaphysical, from the earliest artifacts that have been found. Modern science, dating only from the sixteenth century has become widespread only in the last one hundred years. The founding hope of the scientific method was that the correct understanding of the machinery of nature would reveal its basic metaphysical secrets. Sir Isaac Newton was an alchemist, someone interested in transforming certain elements into other more precious elements such as gold and silver. Pasteur and other major scientists of the nineteenth century believed in "the vital life force," which was presumed to be present in all forms of life.

In less than 500 years the modern scientific method appears to have gone full circle. Einstein died disappointed that he had not achieved an integrated understanding of the mechanical structure of the universe. Leading physicists today, such as Stephen Hawking (who now occupies the same chair once held by Newton), find physics once again verging on the metaphysical.

Isaac Newton, 1642-1727.
[Metropolitan Toronto Library Board]

Twentieth century physics has taught us that our observation of physical phenomena affects what we observe. The classic notion of the objectivity between the observer and that which is being observed breaks down. We know that matter is a form of energy and that matter and energy interact in certain ways that can be described as mathematical probabilities. Somehow the absolute blueprint of the cosmic machine is not possible.

One of the most interesting facets of our armchair adventures in the computer conference which later developed, was the wide variety of ways in which people dealt with the question of the Face. The various methods that people used even to construct the questions posed by the images from Mars ranged from the scientific to the religious, with several variations in between. The overall effect was like walking through *Alice in Wonderland*, gliding through the *The Hitchhiker's Guide to the Galaxy*, or trying to outline the works of Monty Python and Kurt Vonnegut.

Stephen Hawking.

Although methodology is one of the most dreaded and boring courses even for the philosophy major, the importance of the correct method is far from academic. Generally, what happens in everyday living is that a particular method becomes the established way of doing things. No one really questions it very much, whether it is a method for storing items in a warehouse or for conducting cancer research. A few adjustments are made from time to time. Usually though, if you try to change the primary way of doing things you run the risk of finding yourself outside of your job, home, social club, or church.

As a result, the arrival of the Martian images — a large face with nearby pyramids which appeared to have a honeycombed structure — was an event which scientists could easily dismiss or find interesting, depending on the method of analysis they were inclined to use.

Much of the drama in this armchair adventure was the way in which various people dealt with the question and the language they used. Although everyone spoke English, people often talked right past each other. The "Face" presented a special challenge to conventional scientific thinking and our basic knowledge about Mars, the solar system, and intelligent life forms. Consequently, the "Face" on Mars had all the makings of a possible scientific revolution.

Those scientists and thinkers who were looking for a new synthesis between mind and body, spirit and matter, a new synthesis for the age of Aquarius or a new Christian synthesis verifying the literal truth of the Bible, would look to the face as a catalyst in their search. For those interested in reviving national interest in the exploration of outer space, a lost civilization on Mars might dispel the gloom that had been created by the same exploration which had produced the "Face" and had shown Mars to be a vast frozen desert under a dust filled sky.

I must admit that I was impressed by Richard Hoagland's background as an advisor to CBS News and his credits as a reporter and science writer. Although he did not have a college degree or any advanced scientific or technical training, he had a keen mind. As we met and talked in the small living room of his cottage, I discovered that he had anticipated the standard scientific questions which I raised. Clearly, here was a science writer with substantial knowledge, an orthodox scientific orientation, who had come across an interesting question while researching another story. Hoagland had studied DiPietro and Molenaar's monograph while working on another project. He asked what can now be considered a fairly obvious question: If the "Face" was built as a monument, where did the builders *live?* Nearby were some pyramids which looked a little the worse after thousands of years of deferred maintenance. Mars has other more famous pyramids on the other side of the planet, the "pyramids of Elysium" (which are three sided and very irregular) and which are more obviously the work of Martian geology. However, the combination of the "Face" *and* the nearby

This view of the Martian surface as televised by Viking dashed scientific and popular hopes of an Earth-like planet. [*NASA*]

pyramids was intriguing. As a freak of nature alone the formation was fascinating. If it was not a freak of nature, the "Face" on Mars could have freakish consequences for humanity's concept of itself, its origins, and purpose.

My first question concerned the plausibility of the image, and whether the natural history of Mars could include a sequence which might at one time have supported life. He had given the matter some thought and we talked about the history of the solar system and the developmental periods on both Earth and Mars.

Richard Hoagland relayed the current scientific consensus that conditions on Mars had been relatively benign in its early period, when the Sun was 30% dimmer than it is now. The presence of uranite crystals (which form in the presence of water) in the oldest rocks on Earth, combined with the evidence of massive flooding on Mars early in its history, has led some to speculate that the large amounts of carbon dioxide present in the early atmospheres of both Earth and Mars created a greenhouse effect which heated the surface of both planets despite the Sun's lower output of energy.

It appears that there was little continuing volcanic activity on Mars and this led to a locking up of vital elements in the crust. The carbon dioxide combined with the water to become rock in the form of carbonates, just as

NASA image showing the Elysium area of Mars and its natural pyramid-shaped land forms.

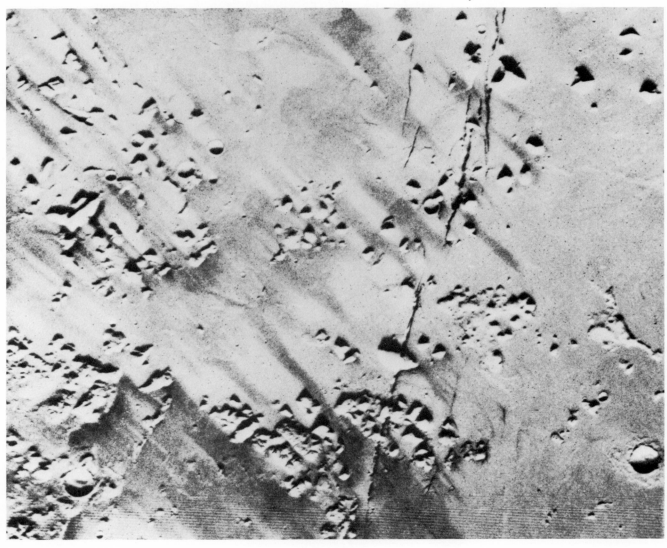

A Profile of Richard C. Hoagland

Richard C. Hoagland's public career has ranged from being appointed (at 18) Curator of Astronomy and Space Science, Springfield Museum of Science, Springfield, Mass., to becoming (at 23) Science Advisor to Walter Cronkite and CBS News. In addition, Hoagland has served as a consultant to several network television projects — from Arthur C. Clarke's 1969 "Spaceward" telecommunications Special, to Ted Turner's Atlanta-based Cable News Network, in 1981. He also served for several years as a consultant to NASA's Goddard Spaceflight Center, working on, among other things, the Agency's Public Service Communications Satellite concept.

At home in the print medium as well as television, Hoagland's most satisfying written contribution to telecommunications was an article on Arthur C. Clarke's historic 1945 "invention" of the communications satellite. Title "A Talking Star of Arthur Clarke," the piece examined the predictions made by Clarke in the light of the implications of the first *direct broadcast* satellite, ATS-6, just launched (in 1974) by NASA. Initially appearing in *American Way Magazine*, the piece was reprinted by the US Information Agency for worldwide distribution.

In addition to communications, Hoagland maintains an avid interest in research — particularly space research. His magazine contributions on this subject have appeared in OMNI, ANALOG and SCIENCE DIGEST. In the mid-Seventies he and several other well-known science writers co-authored a book *Close-up: New Worlds,* St. Martin's Press, 1975), examining current solar system discoveries stemming from NASA's "planetary program."

His own research contributions include: 1) The Pioneer 10 Plaque, Mankind's First Interstellar Message (suggested to Carl Sagan by Hoagland and Eric Burgess, in 1971, and acknowledged in SCIENCE, March 1972); 2) The "Galileo Experiment, carried out by Astronaut David Scott on Apollo 15, in 1971 (the simultaneous dropping of a hammer and feather on the airless lunar surface before a world-wide television audience estimated at half a billion people — to confirm Galileo's prediction regarding how fast objects fall in a vacuum re-

gardless of their mass); 3) The Europa Proposal, that pre-organic material or even simple organisms exist in a satellite-wide ocean under a crust of ice on the second moon of Jupiter. (When initially published — January, 1980, *Star & Sky Magazine* — the concept created a whirlwind of popular reaction, including scores of headlines in the daily press. Scientific reaction was equally strong, ranging from initial outright dismissal to plaudits from Dr. Robert Jastrow, then head of NASA's Goddard Institute for Space Studies. Jastrow termed the Europa Proposal "the first new plausible location for life in the solar system in ten years. . ." Perhaps the most satisfying reaction came, however, from Arthur Clarke, who acknowledged in *2010: Space Odyssey Two*, "This quite brilliant concept. . . may provide one of the best motives for the projected Galileo mission — NASA's 1988 unmanned orbital reconnaissance of Jupiter.")

Hoagland's current research interest lies in verification of his "Intelligence Hypothesis" concerning possible artifacts he discovered and mathematically mapped on several Viking images of Mars. To this end, he has organized two scientific investigations of the material (The Independent Mars Investigation, 1984; and the Mars Investigaion Group, 1986). With the support of former Apollo Scientist-Astronaut Dr. Brian O'Leary, and Presidential Space Commissioner Dr. David Webb, he has also formally created The Mars Project, a non-profit scientific and education research effort to manage the on-going inquiry into the nature of these objects, as well as to inspire the necessary return missions to Mars for the purpose of ultimate verification of the Intelligence Hypothesis. Hoagland currently serves as The Mars Project's President and Executive Director.

Currently finishing a three-year book project detailing the data which substantiates his Mars Intelligence Hypothesis, Hoagland lives, writes and consults from Oakland, California.

it does on Earth. The big difference is that plate tectonics on Earth — the movement of portions of the crust which creates earthquakes and volcanoes — recycles the carbonates, nitrates, and sulfates and releases carbon dioxide and water back into the atmosphere, along with carbon, nitrogen, and sulfur, all of which are needed for living organisms. This lack of volcanic recycling, according to Hoagland, was one of the major factors which turned Mars into a dead planet. Despite the fact that the Sun was growing brighter, Mars was growing colder since its water and carbon dioxide were bound up in carbonates.

He also pointed out some of the controversies to indicate that there were still many unsolved mysteries. One of the most critical was, where did the water go? Were there once large oceans and rivers on Mars? In a few brief words, Richard Hoagland sketched a puzzling paradox. If the scientific consensus about Mars indicated more favorable conditions for life early in the history of the solar system, the same consensus about Earth indicated that life's truly exciting developments are much more recent.

As an anthropologist, I was intrigued, especially since there is still so much controversy about human origins and mythological stories about sky gods and celestial visitors. I wondered whether this might be a replay of Von Daniken's *Chariots of the Gods*, which reinterpreted many ancient ruins and structures on the earth as landing strips or bases for heavenly travelers and their UFO's. Hoagland anticipated my concerns, which served to reassure me, as did his focus on the geology of Mars and the possibility that the ''structures'' might be freaks of nature.

I left our first meeting wondering how one would deal rigorously deal with this type of question. My sketchy knowledge of earth-based geology and archeology told me that one would have to mount an expedition and thoroughly and carefully research the site. How could one responsibly spend millions or billions of dollars on something which might turn out to be a wild goose chase? Kathy and I mulled over these things on the way home. I had told Hoagland that we were going on vacation and we would get together on our return.

Recruiting a Panel

A few weeks later Ren Breck invited us over to his home for what Richard Hoagland would later call the ''Mars Bar-b-que.'' I was intrigued by all of the questions raised by the ''Face'' but I knew that an interdisciplinary team would be required to perform a complete analysis. Due to the lack of scientific reaction to DiPietro and Molenaar's research it was especially important to find out what other scientists thought about the ''Face'' and the idea of a lost civilization on Mars. Richard Hoagland agreed and Ren Breck accepted my suggestion that we begin a computer conference. Notepad had been used previously for mining and construction projects but it had not been used for this type of research. Although we didn't realize it at the time, Notepad would make it possible for us to maintain the traditional type of small group structure involved in research while being able to use expensive equipment and facilities spread across the country. The development of the microcomputer, the modem or computer-telephone connection device, and InfoMedia's Notepad software would allow us to combine the advantages of institutional and individual research.

Until the 1920's scientific research in the West was done primarily by individuals working on their own or with limited personal grants awarded by monarchs, governments, or wealthy individuals. For the most part these individuals were men, although there were a few notable women. It is true that research was carried on in universities and monasteries but the twentieth century research institution which we now depend on is a recent development. Even institutional research was relatively small until its rapid growth after World War II. Until very recently, scientific research was a cottage industry performed by individuals who belonged to various societies or clubs. These scientists would share and evaluate each other's work at meetings and in the publications of the group. In the humanities and the behavioral sciences, and especially in mathematics research, the development of ideas or theories still occurs this way. Thinkers in these fields work alone or in small groups. However, those disciplines such as chemistry, nuclear physics, and medicine which required large amounts of equipment and facilities required the researcher to become the employee of a large institution supervising a large group of technologists.

Although the institutionalizapion of scientific research took place within the university structure of the West, special research centers or think tanks were created which had personnel with no teaching responsibilities. The Stanford Research Institute (which later became SRI International) is one the most highly regarded think tanks, and would come to play a central role in the Mars project. Through a mutual friend, Arthur Young (inventor of the Bell Helicopter), Richard Hoagland met Paul Shay, who was the SRI Vice President for Public Relations. Shay in turn introduced Hoagland to Lambert Dolphin, a senior research physicist at SRI who had done extensive work in Egypt and the Middle East using physical analysis techniques to study pyramids and lost cities. In addition, Dolphin had special expertise in the use of aircraft and satellite images for the discovery and analysis of ancient structures on the planet Earth. Richard Hoagland's connection with Lambert Dolphin and SRI was especially fortuitous due to SRI willingness to explore certain topics which the scientific community at large considered to be slightly (or completely!) unorthodox.

The vast majority of SRI International's work has always been and still is done within the orthodox limits of current scientific method. In fact, much of SRI International's contribution in education and health sciences has been to expand the beneficial applications of twentieth century science. Nevertheless, the current developments in theoretical physics to which we have already alluded have cast some doubt about the relationship between mind and matter. During the sixties and seventies anthropologists studying shamans or ''witchdoctors'' described alternative or psychic ways in which people explained how things happen. The influence of Eastern religions during the same period gave westerners a somewhat more satisfying way of relating to life and the universe than that afforded by western science and religion, which still split mind and body, spirit and matter. Parapsychology developed alongside orthodox psychology in order to deal with psychic experiences and other experiences outside of our normal state of consciousness.

By pushing physics to its present limits, scientific thought had come back to metaphysics. The investigation of parapsychological phenomena became acceptable in a variety of institutional research settings. SRI International received a lot of attention for a project which it performed involving the famous psychic Yuri Geller.

SRI International's willingness to do this type of research was beneficial for the Mars project, but it also raised questions in the mind of orthodox scientists in other institutions about the Mars project. The mere fact that SRI International was interested was enough to start the theme music of Rod Serling's *Twilight Zone* in many an orthodox ear.

Reaction of Researchers from Various Disciplines

In those early months of the project and throughout its history the reaction of most scientists who considered it tended to be, ''Oh, the 'Face' again.'' Others were committed to priority projects and had to evaluate the potential benefit of spending years developing their expertise about Mars. In studying the ways that new ideas and technologies are accepted by various societies, anthropologists and sociologists tend to agree that only certain people in the society, who for one reason or another are slightly outside the mainstream, are open to accepting social or technological innovations. For example, in contemporary American society, computers are more easily accepted by the young than by their parents.

The potential consequences of the questions posed by the ''Face'' made it the type of research problem which the professional scientist could dismiss easily; either the Face was a trick of light and shadow or, if the landform actually was there, it was a freak of nature because the planet never had a developmental phase suitable for an advanced life form. While there might be a few eccentrics who would cast their lot in with the UFO crowd or popular misconceptions such as *Chariots of the Gods*, responsible scientists would stay within the established Cartesian scientific methodology and concentrate on other research questions.

In all fairness to the orthodox scientific establishment, there is another important reason for dismissing the ''Face'' on Mars. The work of scientific researchers is, for the most part, supported by public money and resources. As a community service and national treasury of ideas, technologies, and future businesses, scientific research must not only be inquisitive but prudent and circumspect. The different branches of science which compete for limited public money must be careful not to discredit themselves by appearing frivolous or empty-headed. We didn't get to the Moon by a lot of wild speculation, but with patient and painstaking work by hard-headed researchers and engineers. Since logical thinking requires a careful, limited stacking of assumptions, quirky items like the ''Face'' require conservative explanations unless there is outstanding hard evidence to the contrary. Without this conservative orientation it would be impossible to build a solid coherent body of ideas and information.

The reaction from possible team members as I began my own ''recruitment program'' was surprisingly non-hostile. Several scientists I contacted indicated that they were interested and would like to be kept informed. Since the project was a volunteer effort the participants had to do it ''after hours,'' which made it difficult or impossible for many. In addition, those of us at the core of the developing group did not want to have a very large group, and we wanted to keep the research fairly quiet so it would not be conducted in the glare of the media.

Our task was not an easy one. The topic was controversial and so susceptible to all types of speculation that there was no real way of restraining ourselves from wandering into this forbidden garden. Since the work was

Rene Descartes was a 17th Century mathemetician and philosopher who, among other things, related space and time in the creation of analytical geometry and the calculus. Descartes broke with the established method of his time, which was to appeal to authorities such as Aristotle. Instead, he advocated a rigorous experimental approach which was based on the two types of logical inference, deduction and induction. Deduction is the process of generalizing from a large number of experimental examples and induction is a process of going from a limited number of experimental examples to arrive at broad generalizations. By the use of quantitative, or mathematical, language and reasoning, Descartes ushered in the age of modern science as we know it today. Even though our elaborations of Descartes' scientific method have become more sophisticated in the conduct and interpretation of experiments, the Cartesian method is still the most commonly accepted notion of science. Recent advances in theoretical physics are beginning to challenge the Cartesian model of arriving at scientific knowledge. For the time being, Cartesian science is still the prevailing model or paradigm.

Rene Descartes, 1596-1650.
[Metropolitan Toronto Library Board]

all voluntary there was no centralized control. There was no boss. The result was a lot of free ranging inquiry which pivoted around the technicalities of enhancing and interpreting the images. We were fortunate to add to our group Gene Cordell, a computer programmer and imaging specialist, and John Brandenburg, a physicist. Finally, none of the participants had previously known or worked with each other, except of course for Vince DiPietro and Greg Molenaar.

InfoMedia generously contributed the cost of the computer time and telephone charges. SRI International contributed image development and analysis and many related expenses.

Hemispheres of Mars showing placement of features.

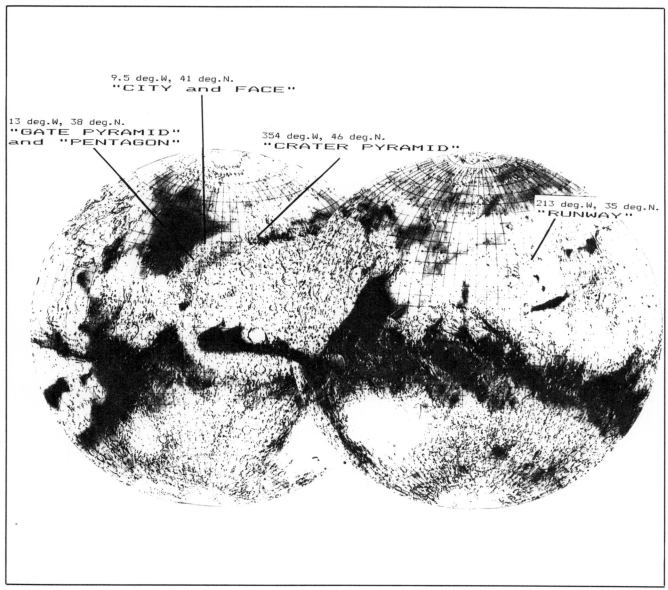

CHAPTER THREE

The Creation of the Conference: Martian Chronicles

A computer conference is very different from most meetings. The design or "architecture" of the conferencing software has a very definite effect on the type of communication available to the participants.

InfoMedia's Notepad is structured to be similar to the types of communication available at regular meetings. Members of a computer conference require a terminal or computer, along with a modem or telephone connection device. Using the modem and the terminal they make a local call

The Notepad User's Guide is a great relief from the usual heavy and difficult software manuals.

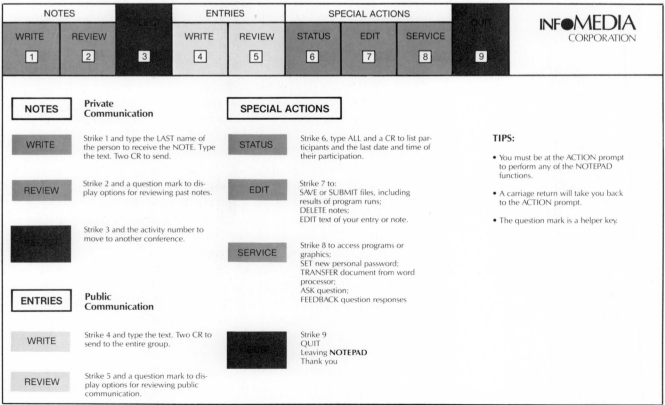

NOTES			ENTRIES		SPECIAL ACTIONS			QUIT	INFOMEDIA
WRITE	REVIEW		WRITE	REVIEW	STATUS	EDIT	SERVICE		CORPORATION
1	2	3	4	5	6	7	8	9	

NOTES — Private Communication

WRITE
Strike 1 and type the LAST name of the person to receive the NOTE. Type the text. Two CR to send.

REVIEW
Strike 2 and a question mark to display options for reviewing past notes.

Strike 3 and the activity number to move to another conference.

ENTRIES — Public Communication

WRITE
Strike 4 and type the text. Two CR to send to the entire group.

REVIEW
Strike 5 and a question mark to display options for reviewing public communication.

SPECIAL ACTIONS

STATUS
Strike 6, type ALL and a CR to list participants and the last date and time of their participation.

EDIT
Strike 7 to:
SAVE or SUBMIT files, including results of program runs;
DELETE notes;
EDIT text of your entry or note.

SERVICE
Strike 8 to access programs or graphics;
SET new personal password;
TRANSFER document from word processor;
ASK question;
FEEDBACK question responses

Strike 9
QUIT
Leaving **NOTEPAD**
Thank you

TIPS:

• You must be at the ACTION prompt to perform any of the NOTEPAD functions.

• A carriage return will take you back to the ACTION prompt.

• The question mark is a helper key.

19

to Tymnet, which is a telephone communications network that allows them to call the InfoMedia host computer in San Bruno, California, near the San Francisco Airport. (This is somewhat analogous to going to a hotel or convention center. When you arrive at the hotel you have to ask for the location of the conference you wish to attend and then register or show your admission papers. Once you are in your conference at the hotel you can be an observer or a participant. If you are a participant you can either communicate individually with the other participants or make a public statement to everyone else in the meeting. When participants dial into the InfoMedia Service Center One they have to submit a series of security passwords once they are registered in the conference or activity.)

Richard Hoagland opened the conference on December 5, 1983 with entry number one at 11:41 P.M. Pacific Standard Time. The conference ultimately spanned the continent, linking Vince DiPietro in Maryland, Greg Molenaar in Minnesota, John Brandenburg in New Mexico, and the other participants in the San Francisco Bay Area.

The Computer Conference

Many people think that science is a neat, orderly endeavor. However, if you have ever seen a scientist's desk or laboratory, you will remember that it is not much different from an artist's studio (or a room in a college dorm!). Doing science is an artform. Scientists, like artists, have their own individual styles and methods of proceeding. Theoretical or ''pure'' scientists often work alone or in small groups in a university setting. Applied scientists and engineers, who sometimes jokingly refer to themselves as ''impure,'' work in crews or teams with elaborate schedules based on deadlines which they had better meet. The participants in the Martian Chronicles computer conference were all independent. Since there was no central funding for the project, no one was in a position to tell anyone what to do or how to proceed. Since the participants were from such diverse backgrounds, each had a slightly different agenda and schedule. However, the research could not really be done by any of the individuals acting alone. Consequently, the project was not tightly organized: mutual persuasion and interest in the subject held the team together for the duration.

The excerpts presented here are far from being a complete transcript of the conference. The conference actually took place in three stages. They represent only the first of what were actually three stages in the discussion, and they cover only four months. The second and third stages of the conference deal much more with speculation about the origin of life, from different theological, philosophical, and scientific viewpoints. Since few of the participants wanted their names to be publicly associated with such radical speculation, these later discussions have not been published in this book. Further, only certain participants gave permission for the inclusion of their conference entries here. Reports on the contributions of the other participants are included in order to convey the context of the dialog. The published paper which summarizes the work of the first stage of the conference has been included at the end to provide a contrast between the untidiness of discovery and the terse objectivity of scientific reports.

In order to keep the story moving, not every entry is included. As a result there are gaps from time to time in the serial numbering. The first entries are dominated by Richard Hoagland's opening of the conference and summary of his research and findings. The various participants also come on stage and take their places.

Face and surrounding landmarks, NASA frame 70A11.

[1] *Hoagland [Richard] 5-Dec-83 11:41PM-PST*

Welcome to a most unusual Activity . . .

An Inquiry into the Possibility of a Former Civilization on Mars

For the benefit of those who may be new to this intriguing tale, here is a brief synopsis:

In June, 1976, the VIKING orbiter 1 snapped a photograph of the surface of Mars, a region at latitude 41 North, longitude 9.5 West. The photograph, some 30 miles on a side, showed what appeared to be a human head — but one measuring a mile wide!

The "object" was dismissed by everyone as a "trick of light and shadow" on an otherwise totally average Martian mesa.

Four years later, in 1980, two independent researchers — Vincent DiPietro and Gregory Molenaar, computer imaging experts working through commercial subcontracts for the Goddard Spaceflight Center — discover the photograph in the National Space Science Data files, and decide to pursue their own investigation into the possible origins of a "human" head on Mars.

For several months they work with copies of the original data tapes, the binary coded recordings of the original images of this region sent back from Mars by the VIKING Orbiter spacecraft. After much analysis of the tapes, much manipulation of the data, called "enhancement" in the industry and identical to that performed on other NASA images by the space agency itself to bring out as much detail as is possible, the two researchers announce their preliminary findings at a June meeting of the American Astronomical Society, in the summer of 1981.

Their conclusion: the "Face on Mars" does not seem to be the product of "totally natural forces. Further investigation seems warranted."

Viking spacecraft. [*NASA*]

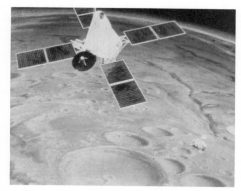

Simulation of Viking landing. [*NASA*]

In the summer of 1983, this author comes into possession of original photos from DiPietro and Molenaar's investigations. After puzzling over the amazing resemblance to a human image on the Martian surface, the author decides to ask the obvious question:

If "the Face" is an artificially-constructed object, ignoring for the moment the enormous problems such a leap requires, then WHERE did the beings who created it stay — while they constructed it?

It is about 10 minutes after asking this somewhat obvious question, that the author discovers what, in fact, may eventually turn out to be the first recognized evidence of the remains of an extraterrestrial civilization . . .

A City on Mars

This Conference is designed to link a hand-picked group of individuals, covering the widest range of scientific and social disciplines, in an effort to determine if this discovery is, indeed, valid. Its secondary purpose, if the validity is established, will be to discuss the implications inherent in this discovery and to suggest profitable directions for further investigation.

It is hoped that a frank and thorough dissection of the data presented to date, as well as any future data, will develop through this effort. Specific comments on all aspects of the information which exists at present, from technical details of the imaging process to the anthropological implications inherent in this discovery — if it is verified — are respectfully requested. Conferees will enjoy full confidentiality in their remarks. Publication of this discussion will take place only after consultation with the authors of specific comments, and only after the First Phase of this analysis has been completed.

It is requested that members of this Activity enter a short professional sketch of their main area of expertise, and how they see this area of interest as assisting with this research.

In entry #30 I state the reasons why studying these landforms on Mars is important even if they turn out to be only interesting lumps created by completely natural processes.

Mission control at NASA.

[30] Pozos [Randy] 3-Jan-84 9:38AM-PST

*The Implications of the Mars Landform Review Project
for the Humanities and the Social Sciences*

The Mars project provides several major opportunities to advance research in the central human questions which underlie the humanities and the social sciences.

Apart from the question of whether the landforms are the product of natural or intelligent processes, the question of other intelligent life forms in the universe raises questions about the nature of consciousness and reflexive intelligence.

Current research in sub-atomic physics, metaphysics, psychology, and anthropology provides new approaches beyond the mind-body Cartesian dualism and indicates that consciousness may be a part of universe extrinsic to individual minds.

The challenge of the Mars project to the anthropologist is to formulate hypotheses as thought experiments that can be tested through enhanced imaging techniques and actual exploration.

For example, there are three central questions:

1. Is reflexive intelligence the product of planetary evolution? In other words, is it independently invented?
2. Is reflexive intelligence something which can be transmitted or diffused?
3. What is the relationship between the physical development of an intelligent species and the development of mind?

Of course there are many additional questions which arise. Even in the absence of another civilization — or remains of one — close by in the solar system or the galaxy, the effort required to examine the plausibility of a society on Mars provides a central crucible for reinvestigating the Eighteenth Century infrastructure of the humanities and social sciences which give rise to our concept of ourselves as a species, our public policy, and the meaning and purpose of our current predicament which pushes us to the edge of extinction.

Whether or not the ruins of a civilization are found on Mars, the project provides a significant opportunity to reformulate the central questions of the humanities and the social sciences. Of course, if a lost civilization is found on Mars the potential consequences are even greater. This will be the topic of another entry.

Your observations and comments are welcome.

Another View of a simulated Viking landing. [NASA]

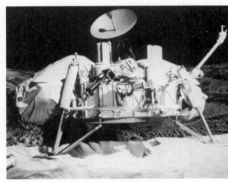

Full view of Viking lander. [NASA]

The Face and the City

After these opening entries, the conference got to work on the cluster of landforms called the "Face and the City."

In a private note which was later introduced into the conference by Richard Hoagland as part of an entry, Lambert Dolphin describes to Vincent DiPietro the location of the "City" relative to the face.

The images are referred to by their special numbering system. For example, 70A11 refers to the 11th frame taken on the 70th orbit of the imaging by orbiter A. The Viking landers were accompanied by two orbiters, with cameras which successfully covered the entire surface of Mars with their mapping photographs.

Pixel The images were not photographs but rather long data streams which described or coded light values for each point in the frame or "picture element." A picture element is technically called a "pixel." Every day we see images made up of pixels on our television screens.

Missing pixels or damaged data comprising a pixel are called "bit hits" and can create the electronic static that we see on television as snow. In imaging, this type of interference is called "noise." One of the major problems in evaluating the scientific plausibility of the "City" and the complete description of the "Face" was the fact that the size of the pixels was too large, 150 feet on each side. If you were to fly over your community at 10,000 feet you could adjust the camera's focus to see everything in great detail down to small cars. However, from a higher altitude, with the same camera you would see only blocks and streets. Most pictures of the Earth from the Moon give very little evidence of our being here. Consequently, much of what we see or fail to see in imaging depends on how much area each pixel or dot in the picture covers. If the pixel covers 100 acres, the level of resolution, we would say, is low compared to images whose pixels covered 1 square foot.

Lambert Dophin refers to the artifact on the camera lens in 35A72. Somehow a tiny drop of moisture found its way onto the lens of the scanner on Orbiter A and created a light, faint bubble image.

Note 182 From Dolphin [Lambert] To DiPietro [Vince] 19-Jan-84 10:24AM-PST

The city can best be seen on full frame 70A11 because of less shadow. Go directly perpendicular to the bright side of the face till you come to the mountains next to the round artifact on the camera lens (35A72). The distance is sort of like 8 miles. The city is a rectangular set of cross hatches at the foot of the nearest mountain. I suppose the individual rooms are about 150 feet across, square, but also with obvious depth. This suggests to me a "honeycomb," as if a bee hive had been opened from the top and a chunk carved out exposing cells at various depths. Light seems to penetrate several levels suggesting a building which was perhaps once enclosed in a dome or perhaps was under a pyramid roof. The streets of the city more or less line up with the orientation of the face and also the edge of the mountain which is triangular shaped. Dick has studied this much more than I so I am asking him to give you more detail. The best we have at the moment is a non-SPIT full frame 70A11 from which we have blown up the area of interest. Will have more prints shortly.

Note 491 became incorporated into public entry number 39 of the conference by Dick Hoagland. This note clearly summarizes Hoagland's discovery of the "City" and Dolphin's role in the analysis.

Note 491 From Dolphin [Lambert] To Hoagland [Richard] 19-Jan-84 11:35AM-PST

Early in December Dick Hoagland briefed me thoroughly on the early work of DiPietro and Molenaar showing a clearly recognizable face on Mars and also a large pyramid, partially damaged on two faces. In addition Dick showed me carefully made enlargements of mountainous regions to the west of the face in which there are additional features not appearing to be natural which DiPietro and Molenaar had apparently not noted before. DiPietro has now kindly supplied much better SPIT processed images of the face and the pyramid showing additional features that are, to me, even more unlikely to be natural in origin. The face by itself or even in combination with the pyramid, impressive as they are, are not as startling as the "city" region in the mountains which is obviously worthy of much more careful study. This I will now do ASAP in order that others on this network may have data for a thorough multidisciplinary study. The city region clearly is something Hoagland's careful study has brought to light which as far as I'm concerned adds much more impetus to the earlier findings. Members of Chronicles wishing prints as they become available should send a note to me or Dick.

Gene Cordell, an imaging specialist at Silvar Lisco at the time of this research, obtained permission to use the facilities of International Imaging Systems, referred to in the entry as "I" squared, "S." This entry becomes the first salvo in the "scan lines controversy." Basically, when the scanner creates the electronic image of a large object, fine lines are created because the image is made up of nothing but rows and rows of pixels. When these strips of coded electronic data are stacked on top of each other in the image fine detail can be obscured or confused with scan lines. The problem as described by Dolphin and Cordell in this entry is that the ''honeycomb'' structure described by Hoagland is small enough in the image to be at the level of resolution. Consequently, since the honeycomb runs parallel to the scan lines, it might be an artifact created by the way the scan lines came together to make the electronic image.

[44] *Dolphin* [*Lambert*] *25-Jan-84 7:43AM-PST*

Gene Cordell and I were able to review the tape recently sent to us by Lynda Sowers of USGS Flagstaff; last night working at International Imaging Systems, Milpitas with Chris Walker. This tape was found to contain 35A70, 35A73, 35A74, and 70A11. No 35A72 though this image was listed on the reel. We paid special attention to the "city area" to the west of the face by looking at 70A11 with filters, all reasonable ranges of contrast and magnification but were unable to recognize any hints of the streets, rectilinear alignments or honeycomb evident in our old 35A72 full frame. In regards to 35A72 Gene notes the honeycomb-like features do, unfortunately, line up parallel with the scan lines and the cell sizes are about the limit of resolution. What is needed next is further examination of 35A72 to make certain we are not dealing with an artifact. SRI senior geologist agrees that pyramids don't usually occur in nature. In addition to the clear four-sided pyramid noted by DiP and M there are several interesting "Starfish" shaped pyramids in our part of Mars. I'll proceed to get 35A72 from Flagstaff and examine the SPIT tapes kindly sent us by DiPietro. (Vince, these are being copied today so originals will come back immediately to you). By the way the image processing capabilities at I-Squared-S are most impressive. I did not see anything on this tape that improved upon the 70mm negatives Vince sent us last week, especially with regard to 70A11.

As an Egyptologist, Dolphin's interest in the pyramidal structures was only logical. One of the early and still unresolved questions in the investigation was the probability of the natural formation of pyramids on

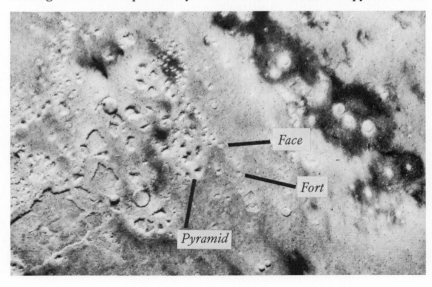

NASA frame 753A33 — the ''morning light'' view of City, Face and Wall [cliff] found by John Brandenburg. This image highlights the other side of the Face.

Bits In entry #45, DiPietro discounts the existence of the ''City,'' based on a technical critique of the composition of the basic dots or pixels which comprise the image. The ''error enhancements'' mentioned serve to introduce two technical concepts of MSB, ''most significant bits'', and LSB, ''least significant bits.''

For those not familiar with computers, the term bits comes from binary digits. Essentially, computers operate by elaborate electronic strings of on and off codes. The number system we use in everyday life is a decimal system which uses 10 numerals. The on-off strings used to operate computers only need two numerals, 0 and 1. In order to describe a character of the alphabet so that the computer using a word processing program can display the letter ''A'' on the screen or print it on paper, 5 to 7 binary digits are needed.

The number of binary digits needed describe each dot or pixel in the Mars images was considerably greater. In the string of binary digits which described a particular dot or pixel, there were ''more'' and ''less'' significant binary digits or bits. For example, if you are trying to find someone's home and the first two digits of the address are wrong, this is a more significant error than having the last two digits wrong. If the address is 3216 you will still land on the right block if the 3 and the 2 are correct. Similarly, if the first digit of a zip code is incorrect then you have sent it to the wrong state. Whereas if the mistake is in the last digit of the new special nine digit zip code, the letter will still end up on the right block, probably even on the right floor of the right building.

Mars. The SRI International geologist referred to is William Beatty. His evaluations will play a special role in the unfolding and direction of the research.

The pyramids of Elysium on the other side of the planet became widely known after their presentation in Carl Sagan's television version of *Cosmos*. Clearly, though, a Face and pyramids brought to the minds of human observers the Sphinx and the pyramids of Egypt. This is an interesting example of the type of projection which occurs in human thinking.

At the end of the entry you may have noticed Dolphin addressing DiPietro in an aside — a type of electronic stage whisper which was not uncommon in the conference.

In a note to Dolphin, DiPietro discusses tapes, images and processing techniques of importance to the conference. Dolphin includes the private note, with DiPietro's permission, as public entry #45. In the first part of the note DiPietro mentions a 1600BPI tape, which refers to the way the data is formatted on the tape in terms of bytes per inch, saying that he may have sent it by mistake. The 1600BPI tape contained major read errors so it was not to be used. The proper tapes were at 800 BPI.

DiPietro has received a copy of the picture of the City sent by Dolphin but believes it includes many error enhancements. He explains that the original data from NASA of image 35A72 included many pixel ''Hits,'' errors which are seen as black or white pixels that are obviously different from the surrounding topography. These errors result when the MOST SIGNIFICANT BITS (MSB) of original data are transposed during their radio transmission through space. The transposition of MSB results in a significant contrast change in the image. During the transmission of 35A72, some of the LEAST SIGNIFICANT BITS (LSB) were also transposed, resulting in subtle changes in contrast in the image.

DiPietro notes that 70A11, on the other hand, contained very few ''Hits,'' at least in part due to the repositioning of the transmitting antenna during the 35 days following the time the earlier image had been received. He recommends checking this later image for the CITY, noting that if it shows up in 70A11, it is there. He agrees that the alignment of X and Y in image 35A72 coincides with the grid layout of the City, noting that SPIT does not ''manufacture'' such lines. He cites his work with Landsat images in support of this statement. Nevertheless, because the Viking data contained so many ''Hits,'' a cluster of them could cause distortion in an enhancement process such as SPIT, which is unable to recognize errors and simply continues integrating the data.

Because of the errors in transmission of data from Viking, DiPietro notes that it is important to ''Clean'' the images before applying SPIT, but he and Molenaar did not have the time and funds to completely prepare them before their analysis. With the raw tapes and the facilities of SRI and other research centers with which the members of the team were associated, he suggests it would be easy to clean the tapes.

He then provides his recipe for cleaning them:

The first step is to do a histogram of the image, removing the extraneous black and white pixels, those with values in the 0-60 And 120-255 ranges respectively. Doing this provides a condensed area of actual data to use.

The second step is to remove all of the errors in each individual frame through an averaging technique. To do this, it is necessary to record the digital values of the pixels around each Target Hit pixel, average those values, and replace the Hit with the Average which has been formed.

The third step is to improve the contrast by stretching the data over the entire range of resolution of the recorder, usually 0-255.

Fourthly, the SPIT should be applied to the resulting data. In this way, the resolution will be improved and the errors (hits) will not be enhanced.

DiPietro closes his note saying he would like to see the results from this kind of test of the process. In the meanwhile, he would like to see Dolphin's prints of 70A11 over the City area to see which features remain visible. He also mentions that he and Molenaar had always been puzzled by an area of the forehead of the Face, above the eye on the shadowed side, which they had nicknamed the Crown. Markings on the images suggested the possibility of ancient construction to them. The Crown is seen in both 35A72-SPIT and 70A11-SPIT. Later, artist Jim Channon will also comment and illustrate the ''Crown'' in his analysis of the images.

Entry #45 is another example of the publication of a private communication. In a face to face meeting communication can often be inhibited because you might not want to embarrass others or to look foolish. Using InfoMedia's Notepad it is easy to communicate freely in private and then to carefully share the communication with others by sending private copies to them or by broadcasting the information in a public entry. For each entry in the Martian chronicles conference there were approximately 5 private notes. The actual number of total notes is not available to the participants during the course of a conference unless the conference organizer discloses it for some reason. Consequently there are often conferences within conferences. Needless to say, this high volume of notes took a lot of storage space in InfoMedia's computer. One possible reason for the large number of private notes may have been that with the exception of DiPietro and Molenaar the participants did not know each other. This lack of familiarity required a lot of getting acquainted time and the development of ideas later presented in entries.

Entry #46 summarizes Hoagland's rationale for hypothesizing a possible lost civilization on Mars or at least indicating the major arguments why this area of Mars deserves another closer look.

[46] *Hoagland* [*Richard*] *25-Jan-84 11:23AM-PST*

Ladies and Gentlemen,

It is time we looked at the "big picture." My original work on the location of "the City" came as a result of applying some basic cultural questions to the problem posed by the existence on the Martian surface of "the Face." If this is, indeed, a monument of "monumental proportions," then someone had to have a reasonable length of time to "carve" it, "sculpt" it, or "construct" it. Even under Martian gravity (.38 Terrestrial Standard) it was a (sorry) "monumental task," easily topping the most impressive artifacts produced on Earth — probably by something like two orders of magnitude. This raised in my mind the obvious question: where did all those people live?! The second factor which drew me toward the region of "the City" was more obvious: the line of the "mouth" of the Face points EXACTLY toward it. This is only another way of saying (after you have — as I did — measured the precise symmetry of this object) that it can best be seen "from the side," as a profile — looking upward toward the sky. There were only two obvious locations from which to see the object in this way: from a suspicious "cliff" perched on a

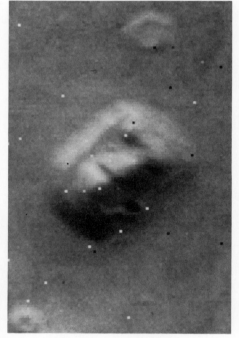

Image of the Face produced directly from data tapes by the prime computer at SRI International.

rampart crater "ejecta flow" to the east (northeast, actually), and from the region of hills and rocks I have since called "the City," to the west (southwest, more precisely).

It was the nature of these two directions — northeast/southwest — which spurred my interest in a possible SOLSTICE ALIGNMENT for the Face and of its best locations for observation. This, in turn, placed my hypothetical "observers" in the center of the complex of hills to the southwest, as only from there could one see the Summer Solstice Sunrise — through the mouth! (And such a sight, I'm sure you would agree, would be MOST impressive....)

And it was after coming to these conclusions through a lot of detailed measurements, and reading of past papers in the literature (Ward, et al.) on the obliquity of Mars, and the changing nature of that obliquity — which dramatically changes the location of the Summer Solstice Sunrise, as seen on the horizon from this latitude — that I arrived at a tentative age for when these artifacts were, indeed, created: no more "recently" than 0.5 million years — by multiples of about a million years, the period of the Martian obliquity changes. It was this reasoning which led me to examine very carefully the strange — nay unique — clumping of "structures" in the complex southwest of the Face, and to (after much internal debate) pronounce it "the City." The overall arrangement of these objects, though huge on a terrestrial scale, maintained a fascinating rectilinear pattern, which assisted my growing belief that this region of the 1000 square miles of Viking frame 35A72 was "different" from the rest. The fact that the eastern "wall" of the first "structure" of the City points EXACTLY toward DiPietro's and Molenaar's pyramid, I took to be significant, as was the precise alignment off true north of all the major "structures" in the City — by an intriguing 24 degrees. The fact that the Face also has this precise alignment off North (24 degrees) was totally consistent with the needs of my (again) hypothetical observers — to view it from the side, for full effect.

The details of this "city" remained strange and unexplained. There were several groupings of much smaller objects which (to an engineer I showed the pictures) seemed too well-placed to be mere "rocks" or "hills."

And, as I pieced this scenario together, there was the consistent and nagging question: why did a portion of the "ground" between the first "structure" and the second (more pyramidal-shaped) object seem to start abruptly — cutting across everything in a perfectly straight line which terminated a region of bright "hills" and even a dark "crater" by a (seemingly) sharp transition to much lower contrast?

It was only after a week or two of examining the photograph that I discovered the tiny "honeycomb cells" in this region — the one location which made sense architecturally (!) for the remains of some kind of shattered and open structure, preserving the details of its insides. The sharp transition between this region and the objects to the east seemed suddenly explained: by a "veiling of semi-transparent material" overlying the objects underneath — exactly the kind of arrangement one would see from a thousand miles up if looking THROUGH a "mesh" too small to be resolved by the Viking image/SPIT technique!

Then there were the details of this "honeycomb." The highest part, in fact ALL the structural details, seemed to make architectural and engineering sense. The highest levels (as judged by the shadows draped across this region from the pyramid-shaped object to the west) confirmed that the region which seemed central to the architecture were highest. They were also ON AN EXACT CENTERLINE with the VERY strange object to the east — which I have termed "the Fort" (because of the easternmost "walls" and "moat" lying inside — which looks extremely "square").

It was only after viewing... and reviewing... countless times... these features, that I perceived that the "honeycomb" seemed to continue east — at one time apparently "overlying" the curious central features of "the Fort." It was Lambert Dolphin at SRI who also — independently — called this to my attention.

The central features of this "honeycomb" are aligned, it is true, along and at right angles to the scan lines of its discovery frame — 35A72. But there are probably as many features of its detail which are NOT aligned, but which form 60 degree, 45 degree, 30 degree, etc., angles to the basic pixel pattern. And, covering the "Fort" there are other, suspicious, "sudden terminations," as if some overlying layer were "veiling" details we are seeing THROUGH another, finer "mesh."

This is why I have deemed it essential to procure the original negative from which this original print was made, as well as to independently create a separate record from our own library of Flagstaff tapes.

It was after viewing these details, of seeing that there was far more to this site than had ever been published, consistent with the presence of INTELLIGENCE, that I began to cast a "wider net" around this entire Martian region.

On the VIKING frames sent to me by Mike Carr during the summer, which spanned the entire 5000 square mile region in this area, I found many other features which I felt were consistent with the remains of a greater cultural complex in this region, capable of supporting the several hundred thousand people who would have been necessary to construct "the Face," to say nothing of the other structures. On the USGS maps I was able to determine that the unique "eroded" nature of the mesas and hills in this region seem to lie on the edges of an ancient ocean— although Carr et al. argue strongly against such an ocean ever filled with water. The fact is, the site of "the City" and "the Face" lies in a relatively narrow band along the ancient "seashore," indicative (to me) that the basic "building blocks" for later settlements, etc., were formed by the unique geology of this particular region of the planet — answering to a first order the haunting question, "Why 41 North, 10 West?"

Thus I come back to my basic premise in organizing this multi-disciplinary investigation: we must observe the interlocking logic of this picture, balancing arguments about "scan lines" against the extraordinary coincidences evidenced by facial features, pyramids, celestial alignments that "happen" to coincide with the earliest dawnings of the human experiment on Earth(!), etc.

There is a bigger picture here. Our purpose should not be to "prove" there was once life on Mars, merely to make so strong the circumstantial case that "something very strange happened there" that we go back — with the technology which can answer once and for all these fascinating questions. And we have less than two years — until the next Martian "window" opens. My objective is to have built a "Grand Jury case" that will permit the sending the Galileo II hardware back to an orbit of the planet, for such detail as is required to resolve these "pyramiding coincidences."

Enhancing the Image: The Honeycomb Controversy

In Entries Number 47 and 48 Gene Cordell states the case against the use of the SPIT enhanced images. Cordell uses a more conservative methodology and indicates that much of the detail which others claim to have seen are artifacts of the image processing and enhancement techniques. These entries appear to ignore Hoagland's call for discussion of the "bigger picture" in the previous entry. This is an example of some of the discontinuities that appear in asynchronous conferencing — that is conferencing when everyone is not present at the same time. A participant may have something special to communicate when he or she enters the conference. This may or may not directly relate to the previous entry. As we proceed through the conference, this "loop de loop" pattern of response will become more evident.

DiPietro and Molenaar's Starburst Pixel Interleaving Technique is a process which fills or corrects questionable pixel values by dividing each pixel into nine new pixels and assigning a value to each through a complex voting procedure. Cordell and Walker used median values instead of using a complex averaging of surrounding pixel values analogous to the "voting" which occurs in the SPIT process.

An average of several numbers or values is obtained by adding them all up and dividing by the number of numbers which were to be averaged. An average is also called an arithmetic mean. A median, on the other hand, is the number or value which is in the middle of the range of values. In other words a median value has an equal number of values above and below it. The difference between the average or mean and the median or middle value can be very small or very large depending on the range of values under consideration.

[47] *Cordell [Gene] 26-Jan-84 1:06AM-PST*

There is much to be said about these Mars images. I am uncertain where to begin. I do have in my possession a digital tape of frames 35A72 and 70A13. These were given me by Dick Hoagland who got them from Vince (I believe). My apologies to Lambert. The frame 35A72 may already have been at IIS but according to Chris Walker they archived it due to its poor quality. I did not realize at the time that it was the missing frame.

Mr. Walker and myself spent several hours examining this image. The method we used is as follows (compare to Vince DiPietro's suggested method):

1. We performed a full frame Median Filter (this is similiar to the averaging technique. It involves summing a neighborhood of pixels and finding the median value, rather than the average. We performed this using many different neighborhoods, e.g. 1 in each direction, to every other pixel to 5 pixels away). This eliminated the noise and HITS (pixel drop outs).

2. We then performed a piecewise contrast stretch. This involves an interactive histogram alteration. With the histogram before us, we dynamically altered the output image to yield the maximum contrast. (This step is principally performed in the IIS image processing hardware and is performed in real time, allowing us to select the output contrast yielding the best image).

3. We then performed a bilinear zoom on the image. This is another technique, similar to SPIT, for yielding a magnified image. I will leave for later a discussion of the pros and cons of the SPIT technique.

We examined the FACE and the CITY in the resulting image. There were no honeycomb-like structures anywhere on the image. We examined it in close detail and observed interesting Martian landscape, and we found the area wherein the CITY should lie, but NO honeycomb structures were present anywhere.

Now to the question of the SPIT processing technique. There are some who believe that it can bring out detail in a picture that was not apparent in the raw data. This is mistaken. It is a fairly good anti-aliasing technique, however it cannot perform miracles. One of the laws of image processing is that you must have a resolution greater than the object one wishes to observe. This does not have to do with processing techniques, but rather with signal and noise considerations. In the examples mentioned in earlier entries, the objects that underwent considerable improvement with the SPIT technique

Further enlargement of the Face, NASA frame 70A13.

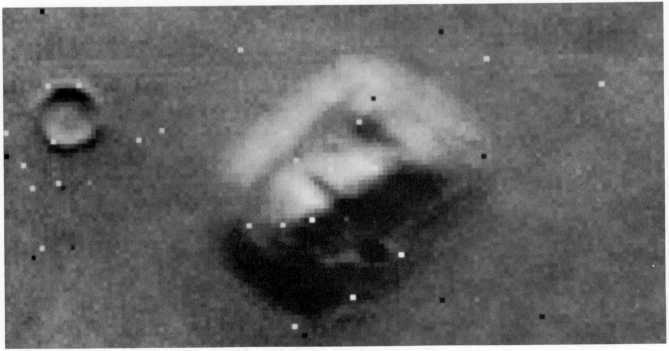

were many pixels in size. The edges were smoothed out, and to the human eye became more recognizable. Techniques for performing this type of operation are well known and are known as anti-aliasing techniques. This is a quite valid application of the SPIT technique. It is not valid to deduce from this or the examples that one can use any image processing algorithm to create greater resolution than was contained in the original image.

It is the case that the honeycomb structures found in the SPIT processed image of the "CITY" are artifacts of the processing.

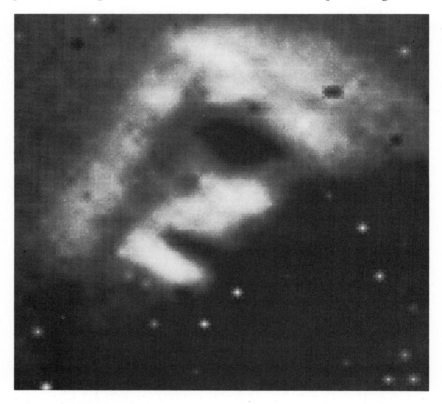

Blowup of NASA frame 35A72, low sun angle.

[48] *Cordell [Gene] 26-Jan-84 2:00AM-PST*

In regards to the face and the other interesting formations nearby I would like to add this. In all of the images I have seen, the Martian landscape is all of a type. That is to say, that we see the same type of geological structure evident in all the frames. There are many plateaus with mounds. In examining these images we found many "faces", some small, some larger, none as large as the one known herein as the "FACE". We found many mounds that could not be interpreted as a face, but which bore the same geological structure. We found many mountains with angular edges. With a bit of imagination, we could come up with many scenarios of incredible proportion for the occurrence of this geology.

What we are dealing with is another planet about the geology of which we know very little. It is quite evident that it is not Earth-like. I believe we should send another space craft to Mars and, in general, continue the exploration of our solar system. It is possible that there is life on Mars. The experiments contained in the Mars lander that were supposed to test for life gave unusual results. They demonstrated that our assumptions about the planet were wrong, and the tests for life were invalid. The question has not been answered.

There is much about the planet that begs for further exploration. There may yet be evidence of other life forms there. Unfortunately, the "FACE" is not an example of this, but a simple permutation of a consistently manifest geological formation.

[49] *Hoagland [Richard] 26-Jan-84 4:24PM-PST*

Let's try another approach:

The area of each Viking frame approaches 1000 miles square. Thus, for any random process, the odds against finding a 1 mile square which contains evidence of that process must be "one in a thousand." Correct?

What are the statistics, then, for locating a pronounced irregularity in the pixels about 1 mile square, lying in the PRECISE 1 square mile most meaningful — in terms of a particular paradigm, i.e. that THAT square mile contains a feature of "significance?"

I didn't find the honeycomb, then decide that it was within "the City." I deduced the presence of the City based on the alignment of the Face — then discovered the honeycomb, in exactly the one location (on the 1000 square miles of Viking data) which makes the most sense in the context of an artificial origin for the objects in question. Can we produce statistics on this?

[50] *Hoagland [Richard] 26-Jan-84 4:32PM-PST*

On Gene's analysis of 35A72:

The fact that a negative (and print) exist which shows the honeycomb within this 1 square mile region of the City, and reveals no similar apparent "structures" (to anywhere near the degree of pefection, regularity, or engineering "sense") elsewhere on this frame, says something. Since the frame I have was not "cosmetically cleaned up" and that is the one which shows this feature, I immediately suspect that "cleaning up the data" also involves "throwing some of it away!"

The region of the honeycomb is somewhat remarkable, incidentally, in its ABSENCE of "bit hits" and noise. There are only one or two within that one mile square. The shadow geometry (from the "pyramid" to the immediate west) fully supports a three-dimensional structure in this area. The contention of you guys seems to be: there's no "honeycomb effect" that you can see, after you "clean up" the data.

Thus, I suspect we are losing subtle data somehow — for the other reasons cited in Entry 46 and 49.

It's fascinating to me that such a crucial question, regarding the actual existence of artificial handiwork in this one frame, comes down to a range of analyses — not merely one!

In fact, it has that central feature in common with much of anthropology and archeology on Earth; there are Great Debates raging all over the place, which are only "resolved" by application of a "combined approach," i.e. economics, geology, myth systems analysis, etc.

I come back to my central question: how did the "imaging artifact" KNOW where to place itself in the 1 mile square most meaningful to an artificial interpretation of its presence?

For, despite your very commendable and thorough job, Gene, you have not yet addressed my other problem: why the surface in this region exhibits other, anomalous and abrupt transitions, "contacts" which argue for a "veiling" of the surface by something we are seeing through? And, in regions of similar contrast elsewhere in the frame, no such "honeycomb" was produced.

Why?

In an effort to facilitate the conference by broadening the scope of the discussion beyond the "honeycomb" controversy I made the following suggestion.

[51] *Pozos [Randy] 26-Jan-84 6:04PM-PST*

I must commend all of you on the caliber of the discussion to this point. However, before we get further enmeshed in the honeycomb controversy, it will be important to compare and contrast imaging methodologies, general features of Martian geology and natural history. In general, I recommend that we spend some time presenting hypotheses from our different disciplinary perspectives and outlining their plausibility as a matter of the conference record before we plunge further into the controversies of prioritizing the likelihood of these hypotheses.

NASA frame 35A72 — SPIT — the Honeycomb, enlarged about 100 times, lies between the Fort and the Main Pyramid of the City.

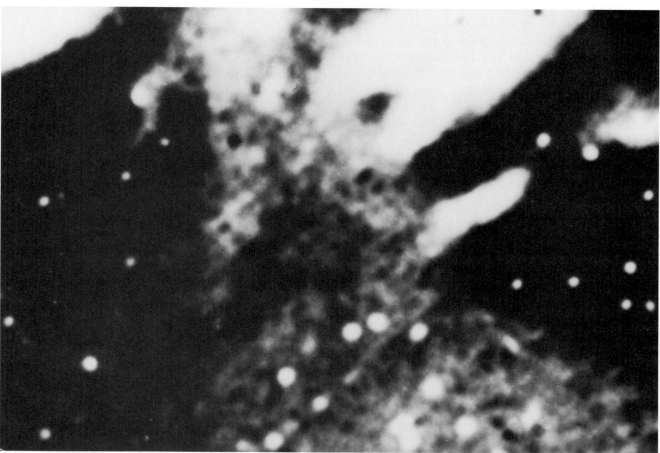

CHAPTER FOUR

Deduction vs. Induction

[52] *Pozos [Randy] 26-Jan-84 6:09PM-PST*

At this point, it would be good if Greg Molenaar and Vince DiPietro could address the characteristics of the SPIT process and discuss their evaluation of Gene Cordell's procedure. Any ideas which you, Lambert, might have based on your experience with ancient structures in the Egyptian desert would be most welcome.

In the next entry, I tried to take my own advice and to present something from anthropology. Unfortunately I used a lot of jargon words which the other participants did not understand. The point I was trying to make was that the experience which we bring to analyzing anything on another planet is earth-based or geocentric. The ways in which we structure information and images in our mind is all based on the structure and function of the human mind. Thus our knowledge is anthropocentric, centered on the human experience. The challenge in doing this type of research is to go beyond these limitations.

[53] *Pozos [Randy] 26-Jan-84 6:14PM-PST*

Anthropologically, the proper analysis of these Martian landforms challenges our ability to reach beyond the conceptual limitations of our species. Generally, our criteria for evaluating — let alone the perceptual structuring — of these landforms and other data sets is based on anthropomorphic criteria developed on one planet.

As humans we are startled by a "human" face on an alien planet. Threatened by the potential self extinction of our species we are further intrigued and alarmed by the discovery of these landforms on a planet identified with the god of war in the codex of archetypes underlying many of the myth systems created by our species. There is evidence of "cratering" disproportionate to the surrounding terrain which suggests some sort of violent end. We perceive pyramidal shapes which evoke the largest monumental structures known on our planet which were associated with highly developed forms of intelligence on our own planet but which are still shrouded in mystery.

The more fundamental question arises. How do we construct and/or interpret data sets from the terrestrial planets with a legitimacy which overcomes our anthropocentric and geocentric experience?

Lambert Dolphin opted for a different approach. In entry #54 Dolphin argues for the more traditional deductive approach. From this point on three different approaches to the topic would be used. Hoagland, Cordell, DiPietro and Molenaar would concentrate on the City and the Face in Cydonia at 41 North 10 West in terms of image enhancement following an inductive strategy. Lambert Dolphin and John Brandenburg would pursue a deductive strategy of combing the planet for unusual surface features which might or might not be similar to those in Cydonia. As an anthropologist I was pretty much limited to my armchair until more information was in on the natural history of Mars. Consequently, I was able to observe people using our earth-based experience to try to understand something on another planet and in the process to see our projection of our concept of ourselves as human.

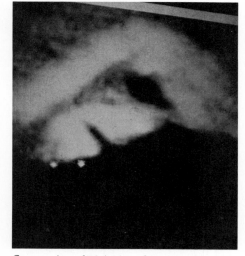

Composite of 70A13 and 35A72-SPIT.

[54] *Dolphin* [*Lambert*] *27-Jan-84 6:37AM-PST*

My approach is somewhat different from that suggested by Randy Pozos in that I feel other examples of interesting landforms on Mars should be gathered and studied prior to formulation of theories about their apparent origin. The more evidence gathered the better the case. If formulation of explanations is delayed, one is usually less likely to be biased by a special theory explaining the features and their associated intelligent life-forms. Therefore I am anxious to track down features on Mars others have noticed but not looked at carefully. I do not personally feel everything hinges on SPIT or

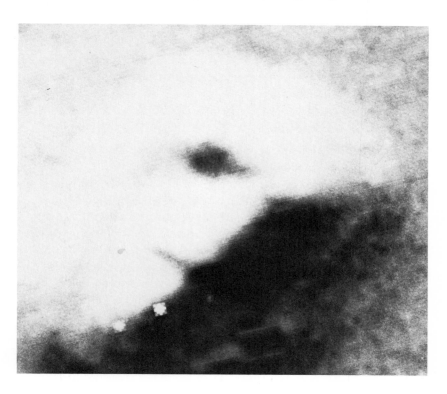

NASA frame 70A13-SPIT showing high sun angle.

alternate processing methods, these are useful when we have things to examine in more detail. I gather for example that many, many images of Mars have never really been inspected closely, even in a cursory fashion? Since the geology is so non-earthlike even more geological studies are likely to come up with lots of surprises even now.

In entry #55, Vince DiPietro discusses both the Face and SPIT processing. He begins with the Face, stating that he does not believe it easy to dismiss it as an typical Martian landform or to compare it to "mounds" resembling faces. He notes that both he and Greg Molenaar have examined many "mounds" and never found as much detail has they have seen in the Face, with its striking resemblance to a human face. The shadowed side of 70A13 is seen to contribute to the impression of symmetricality by duplicating the hairline and eye portion of the illuminated side. In both frames, their examination has revealed what appears to be an eyeball in the cavity on the lighted side. DiPietro expresses himself willing to examine any print of "the many faces" which remotely approximate the feature which is called the Face. He mentions that their inquiries about "faces" on earth have revealed only profiles which had the appearance of faces and cites the Acancagua face in South America as an example of a profile which bears little resemblance to a face but has been cited as having more detail than the Mars Face. In a written joke, he notes, "I remain un con Vinced."

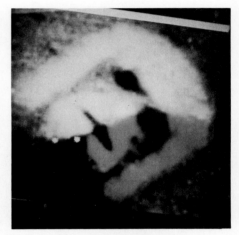

Composite of 70A13 and 35A72 — SPIT, flipped to create an image of the missing half of the Face.

The entry continues with a discussion of SPIT processing in response to Cordell's entries. DiPietro explains that while the SPIT is indeed an anti-aliasing technique, the process employs a digital program rather than using analog smoothing techniques and results in nine times more pixel data for analysis than had been available previously. As a result, the SPIT provides 81 possible new shapes which may be created digitally from any one target pixel of original information. The reliability of the pixels formed in the SPIT is based on use of immediate neighboring pixels in the voting process. He notes that use of the SPIT method on Landsat images revealed the edges of runways and the shape of the Pentagon in Washington, D.C. while other methods they tried produced less satisfactory results.

[58] *Hoagland* [*Richard*] *28-Jan-84 11:31AM-PST*

A couple of thoughts regarding the reality of the Face as a Monument....

I took careful pains this summer, in reviewing DiPietro and Molenaar's monograph, as well as the actual photographs, to attempt a NUMERICAL analysis of the data in hand. If we are dealing with an extraterrestrial civilization, they must have operated on a number system in concert with whatever engineering skills are evidenced in the works we are seeing in the images. This numerical system, therefore, should be evident to the appropriate analysis.

Example:

In determining the orientation of the Face, I had to establish a centerline, and then a measurement of the degree this centerline was shifted off True North. This is how I arrived at the angle: 24 degrees (plus or minus 0.5 degrees). This was strikingly parallel with other angular measurements I conducted on the City, starting with the eastern "wall" that is aimed toward DiPietro and Molenaar's pyramid.

However, as I continued making measurements on and around the Face, I discovered several other interesting "coincidences." If one placed oneself "as an observer" right between the eyes of this impressive figure and looked due South....

The sightline would cross PRECISELY the outer corner of "the mouth." And the angle represented by this sightline (off True North) was once again... 24 degrees!

Such a sightline, incidentally, would take one directly past the eastern edge of a large mesa located to the south — exactly the kind of natural landmark referent which would be essential in "laying out" the proportions and dimensions of such a massive "work of art."

Which brings me to my last point here:

All of the discussions regarding the "reality" or "nonreality" of the Face as an artificial figure have neglected one vital element in any such discussion: there are established mathematical proportions for human images — ratios of distances between the forehead, eyes, nose, mouth, and chin — that are evident in art around the world (at least this one!). What we desperately need is THAT kind of mathematical analysis of this representation of a human face on Mars. And that was one of the key reasons why I (along with Randy) organized this Conference: to get people to contribute that analysis to the on-going discussion of this problem. The same mathematics, incidentally, can be tested in the City — as has been briefly mentioned in an Entry submitted by R. Pozos earlier.

In other words, gentlemen, unless we turn this discussion into a NUMERICAL discussion — measuring quantitatively those things we can measure — we'll continue to just "spin our wheels."

[59] *Dolphin [Lambert] 31-Jan-84 8:58AM-PST*

Lynda Sowers of USGS Flagstaff will check and see if she can get us a tape of 35A72 and also see if she can locate 595A10, 595A11, and 595A12 which Vince DiPietro advised me are images over our area of principal interest. Unfortunately JPL is recataloging Mars data and tapes will not be available from them for six months so we must rely in the meantime on existing data available elsewhere. Anyone wishing digital tapes of other interesting features on Mars should let me know as Lynda is most anxious to help us drawing from tapes she has there.

In entry #60, DiPietro suggests that Dolphin, using DiPietro's name as reference, also request that Lynda Sowers recover a data tape covering a geometrical elongated swath of Mars to which Dr. Snyder of the Jet Propulsion Laboratory had referred at the "Case for Mars" conference several years earlier in Colorado.

Participants were able to log on to the conference from just about anywhere in non-communist countries by making a local telephone call using a modem and a terminal. Richard Hoagland reported from Los Angeles.

[62] *Hoagland [Richard] 2-Feb-84 6:20PM-PST*

On the Road in L.A.... Jose Arguelles reports finding a clipping from 1947 describing a sculptor, Isamu Noguchi, who proposed a "large human head projecting face skywards, the nose to be one mile long ..." The work was to be entitled, "The Sculpture to be Seen from Mars."

H-m-m-m-m-m-m-

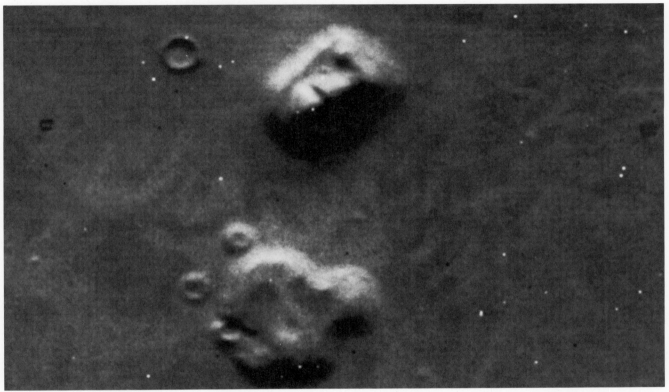

NASA frame 70A13 — SPIT, pulling back to show surrounding surface features.

The Simplest Explanation

In one of the more interesting entries in the conference, Gene Cordell comments on his own entry #48. Cordell doesn't back away from his criticism of Hoagland's hypothetical city but calls into question his own assumptions. By referring to Ockham's razor, Cordell is directing the other participants to search for the simplest logical explanation.

[63] *Cordell [Gene] 5-Feb-84 6:17PM-PST*

It may be a bit presumptuous of me to say this, but did any of you notice that in entry number 48 the author made some very remarkable assumptions?

First, he makes the assumption that because there are many faces apparent on the surface of Mars, they must be natural features. It may very well be that some space-faring race created many of these things for some reason known only to themselves. The fact the Martian surface has many plateaus with mounds would make this an ideal planet for artistry of this type. Taking advantage of the natural "terrain" is a most pragmatic thing to do.

Secondly, he makes the assumption that it is improbable that there is evidence of alien life on Mars, which relates back to the quite common human egoism that pretends that humanity is the only life form in the universe. It may very well be the case that there is evidence of alien life strewn all over our solar system and quite possibly here on Earth as well.

It is incredibly narrow-minded of these psuedo-scientific types to foist these bold assumptions on us unquestioningly.

(Please note that though the above may appear tongue in cheek, these assumptions are made, but by the principle of Ockham's Razor are justified. This principle states that we should make the minimum number of assumptions in order to explain the phenomenon. I wish to point out that it is through the questioning of our assumptions that the most notable advances in science come about.)

In entry #64 Greg Molenaar responded to the controversy by opening his statement with the following words:

"If a mountain clump on Mars looks like a carved humanoid face, the most simple explanation may be that it is!"

Molenaar also raises an interesting question. If the pyramids are naturally formed wouldn't the "rubble" or debris tend to accumulate at the centers of the sides? He also speculates that the debris at the corners of the pyramids may have been the remains of buttresses which reinforced the pyramids. This speculation was later disproved.

Molenaar also appeals to a legal analogy similar to Hoagland's "Grand Jury case" analogy of entry #46, by asking how many points of circumstantial evidence are required for a conviction. Meanwhile Lambert Dolphin and John Brandenburg, a physicist at Sandia Labs in Albuquerque, New Mexico, go image-hunting, pursuing their deductive strategy.

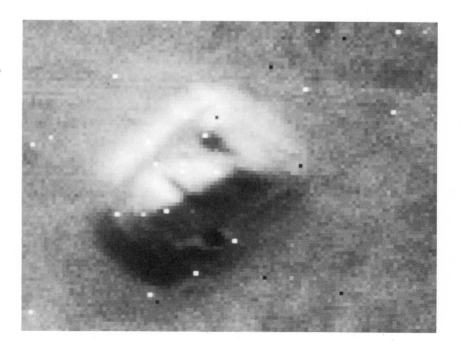

NASA frame 70A13 yet again, this time produced directly from data tapes on the Prime Computer at SRI. Often referred to as the "blue prime image."

Hunting for Faces

[65] *Dolphin [Lambert]* 7-Feb-84 9:25AM-PST

John Brandenburg of Sandia, who will be joining this conference shortly, has obtained some sunrise images of the pyramids of Elysium which he says are interesting. A set is being forwarded to me and I will make copies. He is also on the trail of some of the other images we all would like to see. His interest adds new impetus to our study and I believe his contributions will be most valuable and timely.

[67] *Dolphin [Lambert]* 8-Feb-84 8:31AM-PST

John Brandenburg from Sandia reports discovering a low-resolution morning image which should include the FACE, Number 753A33. He is ordering prints and will advise us on Notepad of results.

In entry #68 Brandenburg announced the exciting discovery of another image of the ''Face'' — 753A33. Brandenburg found the image while searching through the microfilm index published by the Goddard Space Flight Center. He describes the image as a low resolution picture and although it is five times farther away than 70A11, it is a morning shot and could resolve the controversy about the symmetry of the face and might even light the chin of the ''Face.''

Brandenburg also encourages anyone else in the conference with access to the data to check series DAS No. 9162514-584, since Goddard had only sent him the series through 9162519. He states that he has found no other images of the ''Face.''

Brandenburg uses the rest of the entry to describe images he has found of the pyramids of Elysium on the other side of the planet.

This entry was done in capital letters because of the nature of his terminal. Ordinarily in a computer conference, capital letters are used only for emphasis, somewhat like shouting in a personal communication. The style is reminiscent of a telegraph, short, direct, and to the point. As participants became more comfortable and familiar with computer conferencing, their styles tended to become more relaxed and less telegraphic.

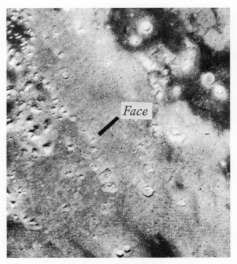

Region of the Face taken from NASA frame 753A33.

[68] *Brandenburg [John]* 9-Feb-84 8:25AM-PST

HAVE FOUND NEW PHOTO OF FACE 753A33 AFTER THOROUGH SEARCH OF MICROFILM INDEX FROM GODDARD LOW RES, 5 TIMES DISTANCE, BUT IS MORNING SHOT AND SHOULD LIGHT OTHER SIDE OF FACE, THIS SHOULD CONFIRM SYMMETRY, ORDERED HIGH RES B CAMERA PHOTOS OF CYDONIA BY MARINER-9(MN-9) DAS NO. 9162514 THRU 584. GODDARD SENT ONLY 9162519 WHICH IS TO SOUTH OF SITE. AM TRYING AGAIN. URGE ANYONE WITH ACCESS TO DATA TO CHECK THIS SERIES FOR POSSIBLE FACE PICTURE AT DIFFERENT LIGHTING THAN PREVIOUS PHOTOS. SHOULD LIGHT CHIN BOTTOM. NO OTHER PHOTOS FOUND OF FACE. NEW PHOTO OF ELYSIUM PYRAMID, HIGH RES B CAMERA MN-9 DAS NO. 12985882 SHOWS LARGE PYRAMID FRAGMENT AT

MORNING, CONFIRMS SHARP FACETING OF SMALLER
OBJECTS. HAVE SENT TO LAMBERT. HAVE FOUND TWO
VIKING PHOTOS OF ELYSIUM PYRAMIDS 732A08 AND
883A03, HAVE ORDERED ALONG WITH 753A33 BUT IT WILL
TAKE TWO WEEKS. COULD SOMEONE LOOK AND REPORT?
ITS GOOD TO BE ON BOARD!

Richard Hoagland seizes on the new image of the ''Face'' discovered by
Brandenburg to pursue his inductive research strategy.

In entry #69 Richard Hoagland provides an interesting commentary on
frame number 753A33. He begins by reminding us that the image is the
thirty-third frame taken by the A orbiter during orbit 753. The orbiters
made one trip around Mars every 24.66 hours at the beginning so that each
orbit would coincide with the length of the Martian day. Consequently,
orbit number 753 occured more than two earth years after the orbiter began
circling the planet on June 19, 1976. The path of the orbiter around the
planet was not a perfect circle. The orbit changed from time to time and
was actually oval (or elliptical) in shape. Consequently, most of the time it
took pictures on the end of its orbit which was closest to Mars. This part of
the orbit is called the periapsis. At other times it took pictures from the far
end of its orbit, called the apoapsis.

In this entry Hoagland infers that Frame 753A33 was taken when the
orbiter was on its way from the apoapsis to the periapsis of its orbit from an
altitude of approximately 5,000 miles. Hoagland deduces this elevation
from the fact that 70A11 and other similar images were taken from
approximately 1,000 miles overhead and that the ''Face'' on 753A33 was
five times smaller. Hoagland also suggests that this image was taken as an
overview or ''context'' shot. He concludes that there may be other similar
images of the ''Face'' in other frames and encourages Brandenburg to keep
looking.

[69] *Hoagland* [*Richard*] *9-Feb-84 6:30PM-PST*

Excellent news regarding additional frames of "the Face." Frame
753A33 is, of course, the 33rd frame taken by the A Orbiter, on orbit
753. This image was taken, therefore, over two Earth years after
orbit insertion of the Viking A Orbiter (one orbit initially was set at
24.66 hours, to be Mars synchronous). Orbit insertion for the A
Orbiter was June 19, 1976. Because of orbital precession and various
orbital maneuvers, the lighting at the periapsis passes changed as the
mission progressed. The data you have given us, John, leads me to
believe that this particular frame was taken from a much higher (5x)
altitude, than the set we've had to study so far — which were all
taken around 1000 miles. This frame would have been taken at
around 5000 miles — probably enroute between the apoapsis and
periapsis, as a "context frame."

My guess is that if there is one "morning shot" like this, there
should be others. The trick will be to locate any that are high-
resolution, equivalent (or better!) than the one's we've already seen.

Good show.... and keep looking.

Lambert Dolphin picks up on Hoagland's previously raised subject —
the return mission. Scientific opinion on the history of water on Mars has

undergone substantial revision in recent years. Dolphin's report cites the growing consensus that water was once abundant on Mars, which seriously increases the chances for life as we know it.

[74] *Dolphin [Lambert] 13-Feb-84 10:19AM-PST*

News item from Aviation Week, 1/30/84: "NASA's fiscal 1985 budget request includes funding for a Mars geochemical/climatology orbiter. The flight is planned for launch to Mars in August 1990. The Mars orbiter is to provide surface chemistry and climate information that, combined with the Viking Mission data, should allow researchers to determine when water on the surface carved the planet's massive canyon features, where the water went and whether that water could ever have stimulated life on the Martian surface."

Security is an ever present concern at InfoMedia. Although the system has never been breached, heavy footsteps from hackers have been heard on the ''roof.'' From time to time the passwords, which function like locks on a building, are changed.

[75] *Pozos [Randy] 13-Feb-84 11:10PM-PST*

DEAR CHRONICLERS:
AS A SECURITY PRECAUTION THE PASSWORD FOR THE GLOBAL ACTIVITY IS BEING CHANGED. TO RECEIVE THE NEW PASSWORD PLEASE CALL:
InfoMedia 415/ 952-4487 Ren Breck or Randy Pozos.
This change will take effect on February 14, 1984. Please change your personal passwords at this time.

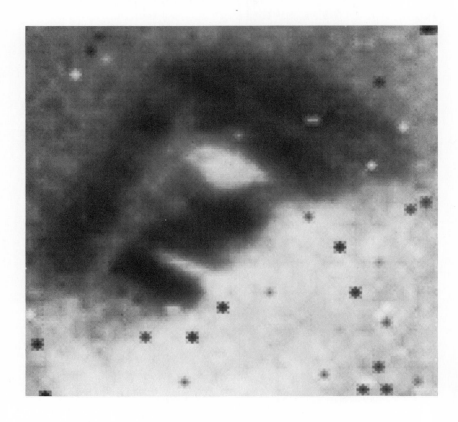

NASA frame 35A72 done as a negative image to enhance the eye and the mouth.

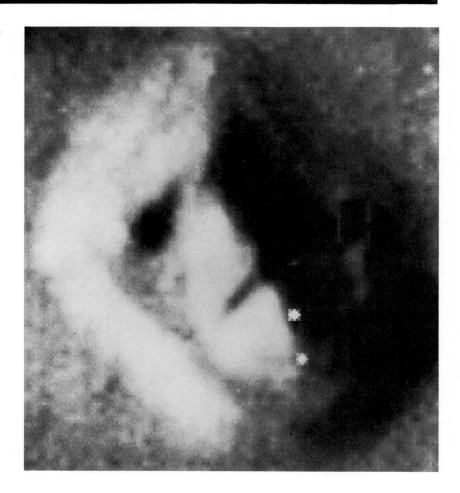

Architecture and Engineering

Further pursuing his inductive strategy, Richard Hoagland elicits the architectural and engineering expertise of an internationally renowned Berkeley architect, on the structures in the "City."

In entry #79 Hoagland reports on the reflections of architect Dan Lieberman, who examined a 16 x 20 image of the "Fort" and the associated "honeycomb." With regard to Greg Molenaar's entry #64, Hoagland reports that Lieberman indicates that a true constructed pyramid would not need buttressing since the entire mass of the pyramid is the "buttress."

However, buttresses at the corners of the pyramids would be necessary if these structures are not true pyramids, but merely as Hoagland reports, "pyramidal shapes hung on some type of frame, which requires an anchoring foundation! Shades of the circus tent idea we had, Lambert."

Lieberman noted that the most interesting feature of the "honeycomb" is not the cell size but the geometric depression in the "ground" just north of the honeycomb. These depressions appear to be very deep, since they have not been filled with detritus, and appear to be artificial. Hoagland was struck by Lieberman's comment that the geometric depressions make engineering sense as part of a structure which was at one time a high technology structure.

Hoagland concludes the entry, "Which brings up again the obvious question: what are we NOT seeing, which a low altitude Galileo II mission could return via CCD television in about 8 minutes (at the speed of light)?''

Lambert Dolphin agrees with Dan Lieberman's assessment that the pyramids of Mars would not require buttressing. In the process, Lambert Dolphin begins another important theme in the conference, the comparison of the pyramids of Earth with those of Mars. These comparisons will be made over several entries spaced throughout this portion of the conference.

[81] *Dolphin* [*Lambert*] *15-Feb-84 8:18AM-PST*

Regarding the pyramids of Egypt, which number more than 70, Kurt Mendelssohn, a British physicist, in his book "The Riddle of the Pyramids" makes a good case that the second pyramid built (at Meidum) collapsed during construction because of steep internal slip planes and inadequate "engineering". The next two pyramids after this were the Red Pyramid (shallower angle) and the Bent Pyramid (angle less steep near top). By the time the best pyramid, that of Cheops, was built the Egyptians knew how to build pyramids so they wouldn't fall down. One feature of the Great Pyramid is that the faces are slightly concave to transfer the weight inwards. The inner (concealed) "steps" of the pyramid are the foundation core masonry, but instead of one steep surface these inner walls are staircased for stability. It never rains in Egypt, but an occasional cloudburst would be enough to have wet the slip-planes at Meidum, causing the collapse. Mendelssohn thinks the workers (perhaps as many as 100,000), were buried beneath the rubble. I concur that the extra material at the corners of the Martian pyramids is not needed for "buttressing" a structure made of stacked stones. Mars being dry we could look into our Mars pyramid for internal structure with an orbiting radar at a high enough frequency to see details. Also the city lends itself to SLAR imaging.

[82] *Dolphin* [*Lambert*] *15-Feb-84 8:32AM-PST*

The photo of the Elysium pyramids supplied to me by John Brandenburg looks to me like 3-sided pyramids, so perhaps we have 3, 4, and 5 sided things out there. I concur that John is the logical focal point and I am happy to copy and distribute new prints from all sources as they become available.

NASA frame 883A04 — natural pyramid-shaped mountains in the Elysium area.

Entry #83 is a continuation of Hoagland's recommendation to detour the Galileo II from its mission to Saturn. According to Hoagland, the $150 million dollars budgeted for the mission included a side-looking radar system (SLAR) for the reconaissance of one of Saturn's moons, Titan. The side-looking radar would allow Galileo II to look beneath the clouds of Titan. Hoagland makes the point that the same system could perform the same task on Mars, provided that the orbit of the spacecraft is close enough to the surface. Hoagland cites Merton Davies of RAND Corporation as saying that Galileo II could maneuver close enough to have a good level of radar resolution of several hundred meters. Anomalous areas could be identified by this type of microwave reflection which also provide limited resolution of images beneath the cover of sand and dust on the surface of Mars.

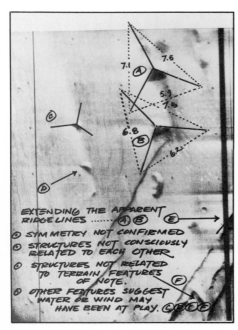

Channon's analysis of Pyramids in Elysium.

EXTENDING THE APPARENT
RIDGE LINES .... Ⓐ Ⓑ → Ⓔ
Ⓞ SYMMETRY NOT CONFIRMED
Ⓞ STRUCTURES NOT CONSCIOUSLY
   RELATED TO EACH OTHER
Ⓞ STRUCTURES NOT RELATED
   TO TERRAIN FEATURES
   OF NOTE. Ⓕ
Ⓞ OTHER FEATURES SUGGEST
   WATER OR WIND MAY
   HAVE BEEN AT PLAY. Ⓒ Ⓓ Ⓔ

Cydonia summer solstice alignments of Face and surrounding landmarks. Small arrow points north.

In Entry #86 Hoagland makes a case for the systematic study of the "Face" and the "City." He criticizes the search for other anomalous surface features as "collecting curiosities," since they cannot be placed in a theoretical framework such as the celestial alignments or the archaeoastronomical orientation which Hoagland presented earlier for the "Face." He claims that this approach will "convince no one of anything!"

Lambert Dolphin holds fast to his deductive approach despite Hoagland's urging.

[87] *Dolphin* [*Lambert*] *17-Feb-84 7:24AM-PST*

In arriving at scientific truth there are generally two approaches. The first is to propose an hypothesis and then look for supporting evidence. The second is to amass evidence and look for patterns or trends from which to deduce laws or probable causes. On Mars I think we are limited to the second method rather than the first as I can think of several hypotheses, all wild and fairly untestable. Therefore the more pieces to the puzzle we collect, however scattered and diverse, the more likely we are to arrive at a reasonable scenario. Also lots of evidence will "convict as well as indict" as John Brandenburg puts it.

Hoagland counters by insisting that there is a hypothesis under consideration, namely his hypothesis about the "City," which he restates in great detail.

CHAPTER FIVE

Celestial Alignments

[88] *Hoagland* [*Richard*] *17-Feb-84 9:59AM-PST*

Lambert, we have a hypothesis "on the table": namely, that an extraterrestrial civilization at some time in the past, when Mars was capable of supporting a non-technological society, developed and built artifacts around the "Prime Site" — 41N., 9.5W. One central "artifact" which they constructed was "the Face," a mile-wide monument of compelling artistry (see forthcoming entry by Jim Channon, a contemporary artist), and equally compelling engineering skill. Other "artifacts" in the immediate vicinity include two five-sided pyramidal-shaped objects — one with three sides of equal length, the remaining two also of equal length to each other, but not to the other three; the other "starfish pyramid" with ALL sides of equal length, capable of being inscribed within a perfect circle.

In addition, the second — more perfect — pyramid sits in the center of a "complex": a collection of larger and smaller objects whose arrangement is anything but random. A "grid" can be placed over the entire 5 by 8 mile area, highlighting a qualitative impression: that the layout of these "structures," including the "starfish pyramid," conforms to a rectilinearity whose most obvious explanation is that it was designed.

Next, there is a striking relationship between this "City" and "the Face." The placement of "the City" is at a perfect right angle to "the Face," facilitating a perfect "profile shot" for an observer located anywhere within the Complex. The line-like "mouth" of this enigmatic object points directly toward the center of "the City." And finally, the observation that this sightline between the City and the Face is oriented southwest/northeast, has led to a testable (!) prediction regarding the artificial nature of all these objects:

Namely, that there was a planned astronomical alignment inherent in their planning, design, and execution on the Martian surface.

The astronomical hypothesis — which demonstrates unequivocally that both the sun and Earth rose over the Face half a million years ago (or 1.5 million years ago, 2.5 million years ago, etc.) provides another crucial test: the general habitability of Mars itself when any of the projected engineering projects were accomplished.

VIKING geological evidence, including images covering the polar caps, speak to a "recent" period (10 million years) of episodic deposition and erosion of material — both water and frozen carbon dioxide snow. The lack of "recent" cratering within this "layered terrain" highlights the fact that this dramatic climate change occurred "recently" in the history of Mars. Climate theories regarding the necessary polar inclination to the Martian orbit predict that more favorable atmospheric pressures will ensue from a higher "angle of obliquity" than the Martian one at present (approximately 25 degrees).

The "astronomical hypothesis" surrounding the "artifacts" at Prime Site I (41N., 9.5W.) predicts that such building would most likely have ONLY taken place following a "favorable" period of climate. Such a favorable period (again, from current papers in the literature of Mars) would only occur when the axis tilt was MORE than the 25 degrees at present, when both poles were more equally warmed by heating from the sun.

The "prediction" seems to fit: the Martian axis tilt which allows the Summer Solstice sunrise to occur over the Face as seen from the center of the City only works for LOWER angles of obliquity — specifically (if the "mouth" is the designated "sightline") a tilt of 17.3 degrees.

In this test, the theory seems to fit!

It is my distinct feeling, ladies and gentlemen, that this kind of careful "mining" of the data at Prime Site I has nowhere near been exhausted. And, further, that it is only via this careful, mathematical analysis, that we will ever make the case for a one-time civilization on Mars.

While reviewing this entry prior to publication, Hoagland asked me to relay his observation that the swings in the axis of Mars (the planet's precession) and the consequent changes in climate are consistent with the supposition of a lower angle of the planet's tilt or "obliquity." From this perspective, the changing tilt of the planet would have provided a favorable climate for the construction of the "Face" and the "City".

This additional observation by Hoagland (two years after the conference) points out another feature of computer conferencing. Even in an "asynchronous" conference, which is a conference in which not everyone is present at the same time, the writer does not always have the ability to draft and redraft a contribution prior to sharing it with the other participants. This feature allows others to be more involved in the development and elaboration of ideas but it also reduces the "tightness" of ideas which we expect in academic or professional meetings.

Dolphin still remains unconvinced and gently questions the fundamental assumption of Hoagland's hypothesis. The hypothesis connot be tested until there is proof that the city exists. The "honeycomb" structure is controversial and far from conclusive.

The Hoagland hypothesis is surely a possibility, but it sounds a bit "Von Daniken" to me until we establish the city's existence beyond doubt. I agree that alignments with the polar axis, sunrise, etc. is good evidence to include and that archaeoastronomy on earth is a valuable new science.

Dolphin's reference to Von Daniken was a decisive intellectual upper hook to Hoagland's attempt to steer the conference. The reference here implies sensational speculation without a reasonable foundation. (Von Daniken was the author of the book *Chariots of the Gods*, which sought to explain many archaeological ruins on Earth as the work of travelers from outer space. Most scientists consider this type of approach to be invalid since there is only theory and conjecture without any substantial evidence.) During the course of the conference and later on several participants were confronted by their colleagues as ''coming close to Von Daniken.''

Meanwhile back at the ''Face,'' the industrious John Brandenburg has found yet another image.

On February 21, John Brandenburg reported the arrival of print 673B56 from the United States Geological Survey (USGS). In this image, the face is smaller than in the high resolution pictures, but it remains visible (7-9 pixels across). The print he received was overexposed but the sun angle was close to that of 70A13 but slightly later and possibly more from the bottom of the picture. Brandenburg expects the data tape to provide more corroborative data to 70A13 regarding the dark side of the face and to provide data confirmng the pyramids and the Fortress. The photo offers good coverage of the entire area. Another photo of low resolution, 9162848 Mariner 9 (M9) A-Camera, also provides broad area coverage. Brandenburg notes that these pictures will be useful for cratering studies. The photo index for Mariner 9 had also arrived and he promised to confirm the presence or absence of high resolution photos taken by Mariner 9.

[95] *Brandenburg [John] 21-Feb-84 5:19PM-PST*

GOOD NEWS: PRINT OF 673B56 ARRIVED FROM USGS. FACE IMAGE IS SMALLER THAN HI RES SHOTS BUT CLEARLY VISIBLE WITH ESTIMATED 7-9 PIXELS ACROSS. UNFORTUNATELY PRINT OF FACE IS OVER EXPOSED BUT AM SENDING TO LAMBERT ASAP. SUN ANGLE IS ALMOST THE SAME AS 70A13 BUT SLIGHTLY LATER AND PERHAPS SLIGHTLY MORE FROM BOTTOM. THIS PHOTO WHEN TAPE IS OBTAINED (AS IT SHOULD DEFINITELY BE) SHOULD GIVE MUCH CORROBORATIVE DATA TO 70A13 AS TO DARK SIDE OF FACE AND ALSO SHOULD GIVE CONFIRMING DATA ON PYRAMIDS AND FORTRESS. I CONSIDER THIS A SUBSTANTIAL ADDITION TO OUR DATA BASE. ALSO PHOTO GIVES GOOD AREA COVERAGE ADDITIONAL PHOTO — VERY LOW RESOLUTION 9162848 M9 A-CAMERA, GIVES BROAD AREA COVERAGE, GOOD FOR CRATERING STUDIES DICK. M9 PHOTO INDEX HAS ARRIVED, WILL CONFIRM WHETHER NO M9 HI RES PHOTOS ARE THERE. MORE LATER.

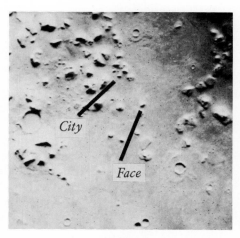

NASA image 673B56, overexposed but slightly different sun angle on City and Face.

An Artist's View

Richard Hoagland, undaunted by Dolphin's reference to Von Daniken, enters the aesthetic judgment of artist Jim Channon on the "Face." (Unfortunately, Channon did not become an active participant.)

[97] *Hoagland [Richard] 21-Feb-84 11:35PM-PST*

The following Entry is submitted by Jim Channon (who will be joining this conference shortly, under his own account).

Jim is a former colonel in the United States Army, assigned to the Pentagon. His current occupation is consultant, mainly in the communications field. He is an accomplished artist, using this talent in furthering communications, particularly in multi-media presentations and corporate affairs.

This, then, is Jim's preliminary analysis of:

THE FACE ON MARS

Three elements will be discussed to highlight my findings after a two-day review of photographs provided by Dick Hoagland.
1. Facial Proportions.... Anthropometry
2. The Supporting Structure.... Architectural Symmetry
3. The Expression.... Artistic/Cultural Focus

FACIAL PROPORTIONS

The artist uses classical proportions and relationships when constructing the human face. The eyes, for example, are only barely above a line separating the upper and lower face. The physical anthropologist relates to a set of classic proportions, that relate facial features in predictable ways.

The features on this Face on Mars fall within conventions established by these two disciplines. I find no facial features that seem to violate classical conventions.

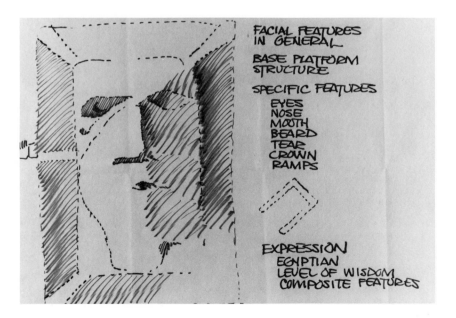

Jim Channon's sketch of the features of the Face.

THE SUPPORTING STRUCTURE

The platform supporting The Face has its own set of classical proportions as well. Were the Face not present, we would still see four sets of parallel lines circumscribing four sloped areas of equal size. Having these four equally proportioned sides at right angles to each other creates a symmetrical geometric rectangle. The photo (70A13) with the 30 degree sun angle reveals that they are clearly formed above the surface of the landscape. These support structure features alone suggest a piece of consciously designed architecture.

THE EXPRESSION

For the artist, there is yet a more precise way to judge the authenticity of this form. The expression expected from one powerful enough to be so memorialized by a monument of this scale would not be random. The artistic, cultural, mythic and spiritual considerations behind such a work of art would demand a predictable expression. The expression of The Face on Mars reflects permanence, presence, strength, and similar characteristics in this range of reverence and respect.

The image appears to be a powerful male of about the right age to be a ruler. Working with materials like stone naturally gives an expression of this size a slightly lifeless quality. That is usually a function of the engineering requirements needed to translate an expression to the grand scale seen here.

But, it must be emphasized that the artistic attention required to generate an expression like the one studied is NOT trivial. Very slight changes in the eyes could create an entirely different kind of character. The shape of each feature in a case like this is a matter of precision.

THIS IS NOT JUST ANOTHER FACE

It is the face of a powerful male character with the strength and age known to have created similar artifacts on Earth.

As an artist with anthropological training (I'll enter my BIO when I log in), the evaluation just presented is overwhelming evidence that the structure revealed in the photographs presented to me by Dick Hoagland is a consciously created monument typical of the archeology left to us by our predecessors. I would need much precise evidence at this point to prove the contrary.

Jim Channon
February, 1984

The City and the Honeycomb

One of the common confusions which occurred in the conference was the identification of the ''City'' with the ''honeycomb'' structure. Although the ''honeycomb'' structure was cited by Hoagland as one of the most significant features of the ''City'' it was not the ''City'' of this hypothesis. In an electronic conference of this type, this sort of misunderstanding is understandable since the participants rely only on written communication. Misunderstandings or miscommunications are more quickly spotted in voice and/or visual communication since the participants can read each other's physical and emotional responses immediately and directly. The advantage of computer conferencing is that it allows the participants more time for thoughtful written responses. Interaction rules in a face to face or voice conference require an immediate response.

[98] *Hoagland [Richard] 22-Feb-84 1:59PM-PST*

I would like to make a crucial clarification for the record.

Several investigators have commented on the existence or non-existence of "the City" in the past few days. It has come to my attention that we are NOT discussing the same features in this discussion!

"My" City is the collection of objects located about 8 miles southwest of the Face, of which there are at least two truly suspicious and very "artificial looking" representatives: the "Fort" (first SE of the Face), and the "starfish pyramid" located immediately to the southwest of "the Fort." Other strange objects belonging to this "City" include a collection of five small objects located in the heart of the complex, a set of small "domes and cones" scattered non-randomly through this small are, and a definite rectilinear arrangement of a small pyramidal-shaped object west of the "starfish" and a long, bright "structure" arranged at right angles to the southeast "wall" of the starfish-shaped object.

The "City" is thus a marvelously arranged collection of large and small features strung over a rectangle of about 5x8 miles. The entire Complex of these "structures" is oriented such that it affords a northeast view of the Face — and of the Summer Solstice circa .5 million years ago.

Blowup of NASA frame 35A72 giving an overview of the City Square, Main Pyramid, Honeycomb Area, and Fort.

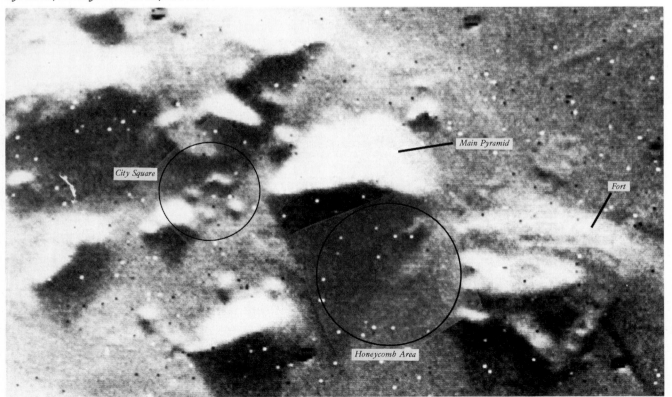

City Square

Main Pyramid

Fort

Honeycomb Area

The "honeycomb" as a sub-feature of the City is a very small object on this scale. It is a DETAIL — not "the" City! Its existence/non-existence seems to have flared into importance because of its OBVIOUS non-natural appearance. My case for the existence of the City is not based on merely the reality of the "honeycomb," but on the overall mathematical arrangement of the entire 5 x 8 mile Complex of large and small objects in this region of frame 35A72.

To repeat:

The "honeycomb" is NOT "the City." The City is a large rectilinear grouping of large and small objects, comprising some very hard-to-explain members, arranged in such a manner as to resemble in eerie detail similar ceremonial complexes on Earth. Its overall relationship to the Face, in terms of pure geometry, would allow the Solstice Sunrise to be seen from the center of this complex circa half a million years ago (if not earlier). The single, predictive question this relationship demands is this:

Was this unique (in terms of other surface features in the area) geometric relationship designed?

[99] *Dolphin [Lambert] 23-Feb-84 10:34AM-PST*

John Brandenburg has received a print of high-pass, morning shot of the face, # 753A33. He is eager to report to everyone that this photo confirms the symmetry of the face and brings out additional detail on the "dark" side. Negative is in the mail to me and also a digital tape from USGS, Flagstaff. I will distribute prints from negative and from tape ASAP.

I am entering this info at his suggestion. He and I do concur with Hoagland's last entry that there are interesting features in the city area (other than the questionable honeycomb) worthy of comment. These features appear in 70A11 and 35A72 so comments from our team on these features is certainly appropriate. Brandenburg is now searching for areas on Mars with interesting features other than our Cydonia region. The face is certainly more interesting than ever now.

Hoagland greeted John Brandenburg's discovery and examination of 753A33 with unrestrained enthusiasm since the new image confirmed the bilateral symmetry of the face — a characteristic of living organisms on Earth and structures made by humans.

[101] *Hoagland [Richard] 23-Feb-84 5:24PM-PST*

On the new images John turned up:
WHE-E-E-E-E-E-!!!! Good show, John.

In entry #102, Brandenburg describes the new image he has just received. The picture was taken at mid-morning, with light coming from the southeast. On image 753A33, the face is about 7-8 pixels wide. The face appears to be generally symmetrical in this image. Brandenburg notes that comparison with 70A13 shows no surprising differences. When he examined the image with a magnifying glass he found the right helmet-face boundary to be symmetrical with the left side of the face seen in 70A13 using falsecolor. He promises to get the picture enlarged as quickly as pos-

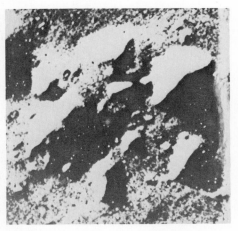
Channon's reconstruction of the City begins with the unretouched image.

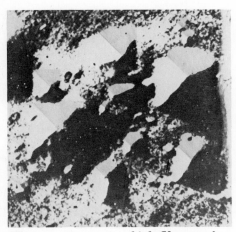
The same image, to which Channon has added partial shading.

The final reconstruction, in which Channon has shaded in the pyramids completely.

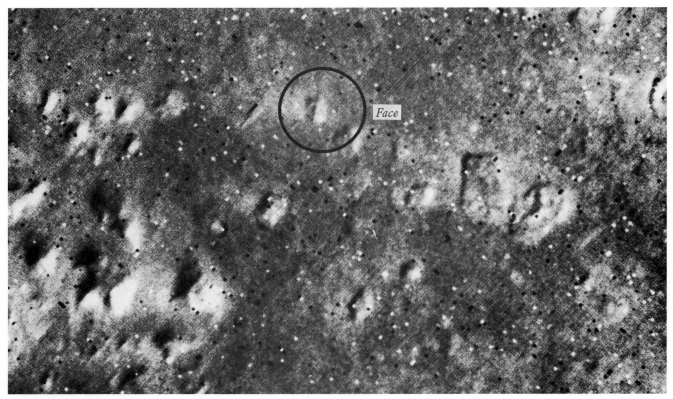

NASA frame 753A33.

sible and states his feeling the the tape of 753A33 will complete the necessary data base for a determination of the artificial nature of the face.

[102] *Brandenburg [John] 23-Feb-84 9:52PM-PST*
EXAMINATION OF 753A33: PICTURE IS AT MID MORNING FACE IS ROUGHLY 7-8 PIXELS WIDE. LIGHT IS FROM SOUTHEAST. OBJECT HAS GOOD OVERALL SYMMETRY IN THIS PICTURE. NO SUPRISES FROM COMPARISON WITH 70A13 CLOSE EXAMINATION WITH MAGNIFYING GLASS CONFIRMS RIGHT HELMET-FACE BOUNDARY TO BE SYMMETRICAL WITH LEFT SIDE AS SEEN IN 70A13 FALSE-COLOR. WILL GET IMAGE ENLARGED ASAP. I FEEL WITH TAPE OF 753A33 DATA BASE ON FACE WILL BE COMPLETE AND GIVE ALL NECESSARY DATA FOR A DETERMINATION OF ITS ARTIFICIALITY WITH HIGH CONFIDENCE. MORE LATER.

In a partial compromise with Lambert Dolphin, Richard Hoagland advocates the review of unusual landforms in the surrounding vicinity of the "Face" and the "City." In the process Hoagland expands his hypothesis to include a broader region of structures around what he hypothesizes to have been a large bay on the edge of an ancient ocean on Mars. The "City" has now grown to become a "metropolitan area."

[103] *Hoagland [Richard] 24-Feb-84 12:51AM-PST*

Regarding additional sites for close examination vis a vis the "artificial structural hypothesis":
On the USGS 1/2,000,000 map of Mars, "Mare Acidalium. . . a controlled photomosaic. . . of the southeast quadrangle" (see Entry 9, 15

Dec, 1984) I can see literally HUNDREDS of features which deserve closer examination via actual prints. In particular, there are some very suspicious objects clustered all along this ancient "seashore," which you should examine in the files, John, if you go to Flagstaff. Before "leaving" the Cydonia region, I urge you to take a close look at these. I have the perception that if the "Martians" utilized existing landforms as architectural units fit for "reshaping," then this region of Mars — located on the edge of this ancient "coast" and amid abundant objects fit for such "reshaping" — would have suited them admirably. Thus, I would suggest getting a copy of this map for yourself (number (I-1351 MC-4SE), before looking at other areas. You may already have examined this region (as you alluded to another "pyramid" located about a hundred miles "down the coast" from Prime Site I), but I thought I'd mention it anyway.

And, in case I haven't said it, excellent work. The more data on the Cydonia region you turn up, the better our case becomes.

Hoagland, in entry #104, then asks Brandenburg for important details about the new frame John has located (753A33) in relation to the ''City.'' Once again, Hoagland draws the focus back toward the ''City.'' The seemingly casual reference to the destruction of the southeast side of the DiPietro and Molenaar pyramid reveals a theme which was to make all of the participants uneasy. The violent destruction of a lost civilization on Mars did not bode well for the future of civilization on Earth.

[104] *Hoagland* [*Richard*] *24-Feb-84 1:11AM-PST*

Can you give us the sunangle of 753A33, the altitude above the surface, and if the frame reaches to the City? The "southeast lighting" you describe should light several of the objects in this region

The Cydonia mosaic image 035A73.
[*NASA*]

from a very desirable angle — including our elusive honeycomb! It should also reveal the destruction of the southeast side of the D&M pyramid in greater detail.

Brandenburg responded at 5:49PM-PST to Hoagland's early morning entry. The sun angle of picture 753A33 was not included in the data block but he estimated it at 45 degrees. The picture was taken at 0933 Marstime. 673B56 was taken at 1725, 35A72 at 1833 and 70A13 at 1650 Marstime. The range for 753A33 was 9300 and for 673B56 was 8400. The frame covers the entire complex under consideration, including the City. A diamond shaped pyramid, the Face and an adjacent object, and another object farther east-southeast of the large pyramid (known as the Citadel). The sides of this diamond have roughly equal sides. There is an object in the center of the diamond at the place in which the lines between the vertices cross. Brandenburg names this object the ''Palace.'' The object next to the head he names the ''Temple.'' The object East-Southeast from the puramid he calls the Citadel. He discovered the pattern because he had to find the head in 753A33 by lining up landmarks. In the process he notices that the head lined up with the pyramid and palace.

Channon's sketch of the mesas.

[107] *Brandenburg* [*John*] *24-Feb-84 5:49PM-PST*

SUNANGLE FOR 753A33 NOT IN DATA BLOCK, ESTIMATE 45 DEGREES. PHOTO WAS TAKEN AT 0933 MARSTIME WHEREAS 673B56 WAS AT 1725, 35A72 WAS AT 1833 AND 70A13 WAS AT 1650, RANGES FOR 753A33 AND 673B56 ARE 9300 AND 8400 RESPECTIVELY. FRAME COVERS ENTIRE COMPLEX INCLUDING CITY.
PATTERN NOTED! — CITY, LARGE PYRAMID, HEAD & ADJACENT OBJECT, AND OBJECT TO EAST-SOUTHEAST OF LARGE PYRAMID (CITADEL) FORM DIAMOND SHAPE WITH ALMOST EQUAL SIDES. OBJECT IN CENTER, WHERE LINES BETWEEN VERTICES CROSS I DUBB THE "PALACE". OBJECT AT E-SE FROM PYRAMID I WILL CALL CITADEL. I NOTICED THIS PATTERN BECAUSE I HAD TO FIND HEAD IN 753A33 BY LINING UP LANDMARKS. DISCOVERED HEAD WAS IN LINE WITH PYRAMID AND PALACE. SORRY WILL HAVE TO REJOIN LATER.

Brandenburg provides additional important details after a humorous reference to Vince DiPietro, ''POOR VINCE IS TEMPORARILY WITHOUT A MODEM, LIKE A COWBOY WITHOUT A HORSE. WE WILL ALL HOPE HE WILL BACK ON THE AIR SOON.'' The analogy that a participant without a modem is like a cowboy without a horse is very appropriate for computer conferencing.
In entry #108, Brandenburg continues his discussion of the pattern at Cydonia. He repeats the description of the pattern in a slightly different format, instructing members of the conference to follow a line between the Head and the D&M Pyramid to cross the object he calls the Palace. A line from the Fort to the Palace will cross another object, the Citadel. Looking at this outline, which is best seen in 35A72, it becomes apparent that the D&M Pyramid, the Fort, the Head, and the Citadel form a parallelogram having almost equal sides. He notes parenthetically that one vertex seems to be half-way between the Head and the nearby object he has called the

Temple. The Palace is at the center of the parallelogram. His reference to "Brothers" near the end of the entry reflects the fact that at this point in the conference, only men were actively participating.

[108] Brandenburg [John] 26-Feb-84 11:22PM-PST

POOR VINCE IS TEMPORARILY WITHOUT A MODEM, LIKE A COWBOY WITHOUT A HORSE. WE WILL ALL HOPE HE WILL BACK ON THE AIR SOON. PATTERN AT CYDONIA — REPEAT: IF YOU GO FROM HEAD TO D&M PYRAMID YOU WILL CROSS AN OBJECT I CALL THE PALACE, IF YOU GO FROM THE FORT TO THE PALACE AND CONTINUE THIS LINE YOU WILL STRIKE AN OBJECT I NOW CALL THE CITADEL. SUDDENLY YOU SEE (BEST IN 35A72) THAT THE D&M PYRAMID THE FORT AND THE HEAD PLUS THE CITADEL FORM A PARALLELOGRAM OF ALMOST EQUAL SIDES (ACTUALLY ONE VERTEX LOOKS TO BE HALF-WAY BETWEEN HEAD AND ADJACENT OBJECT, THE TEMPLE,) SIDES ARE PARALLEL AND PALACE IS AT CENTER, TAKE A LOOK BROTHERS AND RESPOND.

VINCE SAYS GODDARD PHOTOLAB PICTURES NOW DELAYED BY PRIORITY JOB. I AM NOW TRYING TO GET STUFF PURELY THROUGH USGS. GOOD NIGHT, JOHN.

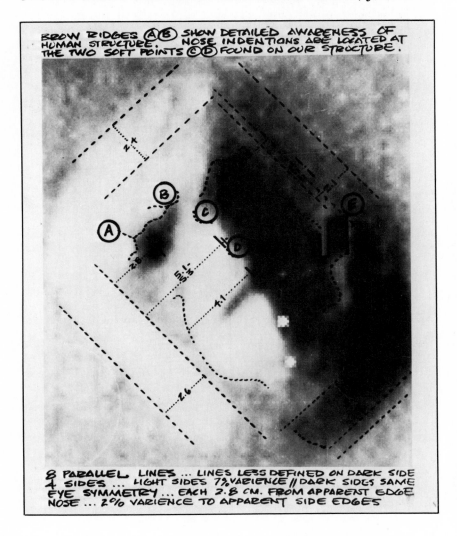

Channon's schematic analysis of the Face based on Hoagland's measurements.

Just a few entries earlier, Hoagland expanded the "City" into a metropolitan area. Now Brandenburg has added a "Palace" and a "Citadel." Clearly, the information gathered by the participants in less than eight weeks began to assume a critical mass. To further the research, outside funding would be needed. However, the only way to generate outside funding would be some type of preliminary report of a conservative nature. Hoagland and Dolphin agreed on the need for publication.

[109] *Dolphin* [*Lambert*] *27-Feb-84 6:32AM-PST*

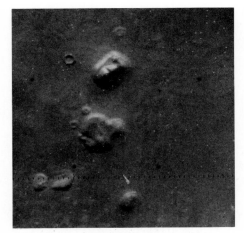

NASA frame T35A13 showing Face.

Dick Hoagland and I feel it is timely to publish a conservative article on artifacts on Mars such as the face, three varieties of pyramids and structures with astronomically significant alignments. Anyone wishing to be quoted or not mentioned at all can simply say so and all will be protected. The article would say such things as the fact that Notepad has made possible an efficient study of evidence by a multi-disciplinary panel of experts scross the country essentially for free (a contracted study like this would already have cost $100k and taken a year). Also, the "distinquished" panel has not yet come to conclusions or made its report, however the evidence is significant enough that quite a number of us are leaning towards the idea that intelligent life once existed on Mars and that further Mars exploration is therefore important. This idea is tossed out for everyone's comments either by entry or private note to Dick. We are thinking of a SCIENCE 84 article, sort of thing. Comments? Thanks.

The actual paper which came about is included in the Appendix. The paper took three months to prepare and was presented the following summer at "The Case for Mars Conference" at the University of Colorado in Boulder.

For the time being Lambert Dolphin has now been drawn into the study of the Cydonia region instead of his wider planetary search. Dolphin's reference to the shroud of Turin which pious tradition holds to be the burial cloth of Jesus was an association which many persons outside of the conference made when they saw images of the "Face." One scientist outside the conference was so frightened when he made the association that he refused to discuss the matter any further since it might lead to scientific support for the validity of religion.

[110] *Dolphin* [*Lambert*] *27-Feb-84 6:53AM-PST*

John, will you sketch up your forts and palaces in the Cydonia area and let me distribute xerox copies to our conferences? Dick has prepared a similar sketch of the city area. Such sketches will help us all talk to one another more intelligently about geometrical features. By the way February LIFE magazine has a portrait of the face on the shroud of Turin prepared by an artist who first made a sculpture from the 3D image of the shroud. Any artist's sketches of the face filling in features that don't require too much imagination would also be of help I think. This is a multi-disciplinary study and I think additional visual aids will be useful to all even though we may later vote down or shoot down various early models or hypotheses. More prints will be available later this week.

CHAPTER SIX

Catastrophe on Earth and Mars

At this point the conference took a definite turn in a new direction. Certain basic geological and climatological conditions are necessary for the development of life as we know it on Earth. Mars, although it is about one half the size of Earth, is very similar in its basic features and composition. Earth and Mars are both classified as ''terrestrial'' planets. However, their natural histories, according to current scientific consensus, have been very different. The prevailing opinion had been that Mars had ''died'' early on in the history of the solar system. Since there had been a great decrease in volcanic activity and no movement of crustal plates as there had been on Earth, it was thought that the necessary chemicals had not been recycled but had been locked up in the crust. This prevented the continued development of the surface of the planet and led to the eventual loss of most of the Martian atmosphere.

If we are to assume that there may have been life on Mars, let alone some type of civilization, there must be some evidence of favorable geological and climatological conditions on Mars at some time in the past. In order to logically support any type of hypothesis about structures on Mars which have survived for long periods of time without being destroyed by weather on the planet's surface, there is a corresponding need to explain the abrupt transformation of the climate and the preservation of the ruins in the Martian deep freeze. The notion of catastrophes in science had been played down for over a century. Evolutionist thinking, following the requirements of Ockham's razor in logic, had sought simple gradualistic explanations of the development of life on Earth. The earlier religious approach had focused on the creation of the universe by God, followed by changes brought about by catastrophes which were acts of God, such as the great flood.

Within the wider scientific community during the course of the conference, catastrophes began to take on a new respectability due to more recent discoveries. Depending on how you want to read the supposed celestial alignments of the ''Face,'' it would be possible to make the circumstantial connection that Mars could have ''died'' at about the same time the dinosaurs died out on earth, making way for the rise of the mammals and eventually the appearance of humans. As a result the supposed Cretaceous Tertiary Event catastrophe was a special boon to the conference.

[115] Dolphin [Lambert] 27-Feb-84 10:03AM-PST

The following general remarks on Mars are supplied by geologist Bill Beatty after reading the references mentioned in the previous entry: Bill personally believes Mars had an early dense atmosphere, storms, weather and warm conditions. He thinks there was then a sudden, chaotic temperature drop and loss of atmosphere. Large quantities of water and CO_2 still exist frozen beneath the surface. Cratering occurred from the beginning. Old craters are sometimes filled with water-borne alluvium or wind carried deposits. More cratering after loss of atmosphere. Volcanic activity continuous since the early days producing literally thousands of craters (as numerous as the impact craters). The crust is thick enough to preclude plate tectonics (as occurred on earth), but when fractured allows magma to surface. Hence the core of M. Bill thinks a Velikovsky type scenario such as a passing close encounter event in the early days. Thus Bill's geological model would be consistent with a climate conducive to life at one time. Sorry about all the extra symbols the computer is introducing today. Call Bill for comments. His Bio was entered last week under my name.

Dolphin refers to the extra symbols which his computer was introducing as he placed the entry. This will sometimes happen in computer conferencing due to minor malfunctions in the modem or static ''noise'' on the telephone lines handling the transmission. Biographies of participants were included in a separate conference ''room'' or activity. Like most conventions in an hotel, computer conferences offer a variety of ''rooms'' or activities.

Building on Natural Formations

Hoagland refers back to Brandenburg's entries #107 and #108 and presents his own perception of the various features as part of larger complex which had been adapted to natural formations but developed according to celestial alignments similar to the development patterns of ancient cities on earth. This is another example of the interweaving of responses that occurs in ''asynchronous'' computer conferencing, that is computer conferencing when the participants are not all present at the same time. (When the participants are all present at the same time it is called ''real time'' computer conferencing.)

[116] Hoagland [Richard] 27-Feb-84 2:15PM-PST

One problem, John, with your "square" is that such a geometrical formation can be made for many such features on this landscape. It is not unique. Furthermore, it is not very precise either. I don't mean to throw cold water on the idea that "they" deliberately constructed such an arrangement, but I don't offhand see why. . . or (more important) how. I tend towards the idea that a lot of the "architecture" is reshaping of what was there — eroded mesas, etc. The particular placement, then, would have been foreordained by geology (with some selection by the later "builders" obviously).

I had the same difficulties with my ideas regarding the "archeo-astronomical alignment theory"; it seems difficult for me to imagine

BUILDING the Face and the City in their respective locations. It seems much more likely that the "builders" took advantage of a pre-existing geometry, and gave it "meaning" by reshaping it, etc., in terms of the Summer/Winter Solstice alignments. A very similar phenomenon took place in the American Southwest, where the local Indians (Anasazi) took advantage of local mesas, outcrops, buttes, etc., in their solar and stellar astronomical predelictions.

Thus, if such a "square" is significant, I feel it is only because it was there before — and the "Martians" took advantage of it (for some obscure reason). I have identified a Solar Alignment for the D&M pyramid (the one on the hill south of "town") with a elliptical mesa to the northeast. The fit (for the same time as the main alignments figured for the City) is EXACT. I was quite intrigued with this, as it was predicted by the alignment theory. I would feel happier with your square, John, if there was an external reason (such as some kind of astronomical alignment) which gave it a reason to exist.

Incidentally, in looking at the mosaic I've made of this entire region (from frames 35A70 through 35A74), I find a fascinating straight line of objects, extending both ways through your "palace." The line continues to the southeast through the corner object in your "square," and continues about an equal distance (about 6 miles) to another mesa. (Intriguingly, if you turn the line at right angles at this point, it runs into another "weird" object — also 6 miles away. Randy Pozos, on looking at this object on the high resolution images Vince furnished, called it "The Burial Mound," as it resembles a large, round breast (!), complete with nipple!!).

Going back to the original line through the "palace": If you extend it northwest, it passes through "the Fort," out across "the lake" (which is what I call the flat, almost featureless plain these objects seem to sit on or beside), until the line intersects another mesa, on the northern "shore."

Then, STILL IN THE SAME LINE, beyond this mesa there is another (VERY interesting) object, about 6 miles further. I've examined this object for quite a while on the "raw" NASA prints Mike Carr furnished me. It seems to have a stepped platform, regular symmetry, and a very curious, pyramid-shaped object "built" on the top of the flat platform.

Now, I have no idea whether this means anything. But the alignment (of surface features) is there.

My personal belief is that, with so many mesas on this landscape, chance alignments — mesas with each other! — is bound to happen. More difficult to explain (I hope) is such an alignment with significant celestial references, especially those that fit the appropriate time-period, in terms of climate that would have made possible a non-technological civilization.

What this is showing us, ladies and gentlemen, is that resolving these questions is going to take a truly multi-disciplinary approach. The engineers have got to get on board, to tell us the amount of work required to reshape versus literally BUILD objects miles on a side! By the same token, we should carefully distinguish apparent order from a true pattern, designed by whomever did all this (note that I'm assuming someone did, in fact, design what we're seeing).

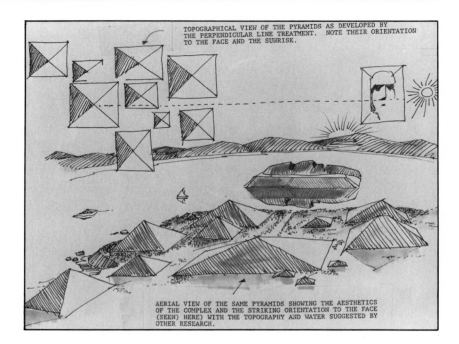

Channon's topographical sketch showing the orientation of the Face to the sun and the arrangement of the Pyramids along the hypothetical shoreline.

I will say one thing, John. Your "palace" lies EXACTLY south of the Face, and forms a perfect sightline for anyone who would have had to "bulldoze or hand-chip away" the stuff on the Face, in terms of the correct "expression" (to use Jim Channon's phrase) or even orientation.

I'll have some further thoughts on this in a separate entry. But, I'm fascinated by how much "projection" it is possible to foist on any problem. As such, this investigation will make a truly vital contribution to fundamental scientific theory, if nothing else! More later.

In the next entry, Bill Beatty follows through on his approach to the natural history of Mars. If, indeed, the planet has been more active and is more "alive" than was once thought, this could account for the unusual formations which the conference participants have been studying.

[117] *Dolphin* [*Lambert*] *27-Feb-84 2:29PM-PST*

Geologist Bill Beatty notes that the face is similar to other "bulges" in the Cydonia area and if one were looking for a natural explanation, he suggests impact craters, slumping and "capricious" wind erosion and deposition. If the face is not natural, then it is evidently carved from an existing hill as opposed to being built up out of blocks. He notes that wind erosion and deposition has been a major factor on the Martian surface. If natural, the pyramids could be remnants or segments of volcanic craters occuring along near-vertical faults in the crust. The Magma pushed up through vertical faults tends to "blockiness", hence angular features.

Gene Cordell picks up on Dolphin and Beatty's reference to planetary catastrophe and inquires about the relationship in terms of time between the extinction of the dinosaurs and a possible timetable for the hypothetical Martian "City."

The End of the Dinosaur

[118] Cordell [Gene] 27-Feb-84 6:46PM-PST

Lambert's entry #115 is very interesting. The dinosaurs became extinct here on Earth between 60 and 90 million years age. One of the latest theories, albeit controversial, is that this was caused by a large object colliding with the Earth, landing in the ocean near what is now Iceland. For those not familiar with this theory, the collision caused catastrophic planetary disruptions and was the cause of Iceland's formation. It threw up so much debris into the atmosphere, that the foodstuffs of the dinosaurs were eliminated for 2 years. The evidence for this lies in the concentration of rare isotopes in a very specific geological layer.

The question then comes to mind, does this time frame correspond to Bill's prospective Velikovsky-type scenario for Mars? As much as I am a total skeptic regarding un-'natural' explanations for the formations on Mars, does this not also correspond to Dick Hoagland's speculations regarding the time frames for "civilization"?

Some research needs to be done on the time frames I mention here; I do not have any references handy with firm dates on the extinction of the dinosaurs or the collision of Earth with a large astronomical object, thus this remains entirely speculative.

Hoagland responds to Cordell's speculation and provides some rough hypothetical timetables. Hoagland sees no compelling relationship between the two events.

[120] Hoagland [Richard] 27-Feb-84 8:03PM-PST

In response to Gene's last Entry: The "Cretaceous/Tertiary Event" (which is what the geologists/paleontologists call the abrupt demise of the dinosaurs — and about 75% of all other life on Earth!) occured about 65 million years ago. It is the theory of Luis Alvarez et al., based on the discovery of a clay layer at the CT boundary strata, enriched in iridium and other noble metals, that an asteroid rich in these materials collided with Earth at that time, causing the extinctions. The exact extinction mechanism, however, is still under intense investigation, although dust — blotting out the sun and resulting in a collapse of the photosynthetic foodchain — has been advanced as a leading concept for the "mechanism," whereby such an asteroid (size:"only" 10km) could have produced such a world-wide catastrophe. The date for this event — 65 million years ago — is quite a bit farther back (in time) than I placed "our Martians." On the other hand, it is not excluded by any of the data in hand, certainly not by either the alignment theory (which works in increments of about 1 million years) or the observed degradation of the various pyramids or the Face itself. The erosion (if that, in fact, is the agent responsible for the bad condition of much of what we observe) is only about a thousandth of a micron per year at present (!), given the present 6 millibar atmosphere (source: J. Pollack, NASA-Ames). This translates to a micron in a thousand years (for rock), to a thousand microns in a million, etc. If we are seeing stuff other than rock, however, this is a very poor way of estimating ages. As is crater counts. Thus, bottom line, I don't see any particular reason for connecting what we are seeing on Mars with the famed "extinction of the dinosaurs," but I don't see any reason not to either. . . at this time. I have additional thoughts, however, on the entire question of "When a Habitable Mars," which I will enter shortly.

CHAPTER SEVEN

Timetables

Interestingly, Richard Hoagland now enters some additional ideas about an hypothetical timetable. Extrapolating from Earth experience, he assumes that pyramids probably represent a low technology civilization — an idea which he will later reverse.

[121] Hoagland [Richard] 28-Feb-84 12:36AM-PST

Some Thoughts on Martian Climate and the Prospects for
A Non-technological Civilization

The discovery of what I have termed "the City" on frame 35A72 raised several important questions in my mind this past summer. The first of which was simply: how could a culture of the apparent non-technological level (on Earth) implied by the astronomical orientations of the face and Pyramids have developed, let alone survived, on the Mars we presently "know."

The immediate answer: it couldn't.

The obvious alternative to the present environment, therefore — if I had discovered the ruins of whomever built (or sculpted) "the Face" — was a much more benign Mars, a Mars which possessed a much denser atmosphere, warmer climate, and — most important — liquid water.

When, in the best theories of the geologists, had that occurred in Martian history? Not for several BILLION years, was the answer — according to experts of no less calibre than Pollack, Ward (of the obliquity question), and several others who have looked hard at this particular problem ever since the Mariner 9 images of Mars returned, in 1971.

But a timescale of several billion years ago for a Martian environment in which the atmosphere was comparable to earth's (in pressure, at least), and in which water was a liquid on the surface, made for great difficulties in trying to explain the presence of a "human" face on Mars — for humans would not appear in the solar system, here on Earth, for several billion years AFTER Mars had "died."

It was a difficult problem.

NASA frame 043A01 showing the Crater Pyramid and Tunnels. These features appear in the Cydonia region 200 miles NE of the Face. The wedge-shaped pyramid at the edge of the crater gives rise to the question: was it created naturally by the same forces that formed the crater, or is it artificial? The crater may be either an impact of volcanic crater, and the valleys in the area may have been formed either by water or lava.

In a private note which Lambert Dolphin entered as entry #122, Dolphin and Beatty make the case for a much more active Mars which did not "die" in the early ages of the solar system. In effect this provides a much wider timetable to allow for the development of life on Mars along with enough millenia to account for the emergence of an intelligent life form. As he does this, Lambert Dolphin plunges the conference into further controversy by attributing the Biblical flood to the destruction of the Earth's vapor canopy by a passing comet.

[122] Dolphin [Lambert] 28-Feb-84 7:09AM-PST

Note 497 From Dolphin [Lambert] to Hoagland [Richard] 28-Feb-84 6:55AM-PST

Bill Beatty does not expecially agree with his learned geologist colleagues about the early history of Mars being over with in billions of years. I don't either. I think this is a result of an unwillingness to see Martian history in terms other than classical earth geology. However on earth the latest thought is now for catastrophes (the dinosaurs at 65 mya (million years ago) for instance). I don't hold to Archbishop Ussher's date for the flood at 4004 BC, I do think there are major surprises in store for us re(garding) Mars. A respected theologian friend of mine upon seeing the Mars photos recently said, "suppose the fall of Satan occurred there?" (Isaiah 14 describes this event). This leans towards C.S.Lewis in "Out of the Silent Planet". An alternate Biblical hypothesis is that men from earth colonized Mars before the flood of Noah, (since the antediluvians were thought to be both numerous and highly advanced). Such an hypothesis would allow a sudden disaster on Mars and Earth at the same time. Until a week ago I never had seen a good mathematical model of the vapor canopy, uniform, pre-flood climate on earth and stable conditions triggered by volcanic eruptions. However such a model is now given by Dillow in his book "The Waters Above". His theory also accounts for the iridium layer from excessive volcanic ash fallout. No particular date for the flood is necessary, it could have been very ancient. The vapor-canopy atmoshpere was very stable, hence a passing comet might have been the triggering mechanism on earth, likewise disrupting Mars. The civilization on Mars might not have

NASA frame 043A03 showing the Crater Pyramid and its Shadow, upper right corner.

died out instantly but over a long enough period to allow shelters and large monuments intended to signal to the earth their presence.

Dolphin's speculation about a dying civilization on Mars is reminiscent of Percival Lowell's speculation that the canals on Mars had been built by a civilization trying to bring water from the Martian Poles to maintain itself. By adopting a position midway between the scientific creationists who claim that the Earth is only a few thousand years old and the orthodox scientific notion which discounts the Bible as a source of scientific information, Dolphin steps into the middle of a growing controversy. Early in the conference several participants reported strong emotional reactions from

Blowup of Crater Pyramid NASA frame A43A04. Notice the large adjacent crater and its ''wormholes,'' parallel rills possibly formed by lava.

"face" and being told that the image was from the surface of Mars. Mainline and evangelical Protestants tended to be among those who reacted strongly by trying to find some relationship between a possible civilization on Mars and the Bible. Dolphin's theological speculations made several of the participants uncomfortable. However, Dolphin and to a more limited extent, Brandenburg began to anticipate the potential social and religious impact this type of research and speculation might have on mainstream American society. Once again, Dolphin has taken the conference in another important and controversial direction.

Meanwhile, back in Cydonia, Hoagland continues his elaboration on the geological and climatological timetable.

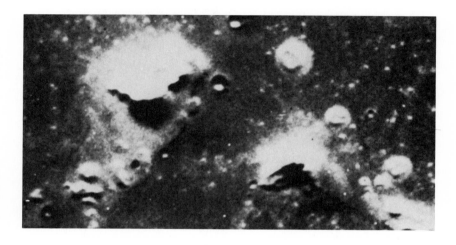

Blowup of NASA frame 072A04 showing the Amphitheater. This surface feature some distance SW of the Face is a remnant of an ancient cratered plateau. Rising 600 meters high, it may be a crater rim. [SRI]

[123] *Hoagland* [Richard] 28-Feb-84 11:18 AM-PST
Some Thoughts on Martian Climate [cont'd]

There are two primary reasons for assuming that Mars "died" several billion years ago: the presence of very old, very cratered "highlands" in the southern hemisphere of Mars; and independent calculations as to what HAD TO HAVE HAPPENED to the carbon dioxide atmosphere originally outgassed by the planet — in the absence of any plate tectonic recycling mechanism.

The "ancient highlands" are thought to be ancient because large crater counts (that is, counts of large craters) tally very closely to similar counts made on other planetary bodies in the solar system — such as the Moon. And for the Moon the actual ages (as dated via various isotope mechanisms) point to a very early event of massive cratering — called the Heavy Bombardment Period — in solar system history. The mechanism is fairly straightforward: the "mopping up" of the interplanetary debris left over from the formation of the solar system itself. This, then, ties such a Heavy Bombardment to about 4 billion years BP (before present), and seems nicely matched by similar episodes all over the solar system, as recorded by a variety of unmanned spacecraft we've sent looking, out to Saturn at this writing.

Since such cratered, ancient terrain is still visible on Mars, almost uneroded (!), it stands to reason that nothing much must have happened in the intervening 4 billion years to wipe those ancient craters away — such as the effects of a heavy, dense, and "Earthlike" atmosphere. Our atmosphere can do phenomenal damage through erosion in a very short period of time (about thousands of years — look at the Egyptian pyramids, etc.), so any similar extensive period for such an atmosphere on Mars (so say the "learned geologists") would have caused significant erosion and modification ALL OVER THE PLANET. This seems not to have happened. And this is explained as the result of whatever dense atmosphere Mars once may have had rapidly leaving — either into space or by combining with the very ground.

Which brings us to point two:

Calculations indicate that carbon dioxide (the predominent constituent of such a primitive atmosphere) would "rapidly" (within a few million years) combine with surface materials on Mars to become carbonates. In the presence of freely flowing water such a process would be even faster (as is evident on Earth, with our vast oceans and hydrological cycle). In a declining epoch of volcanic activity (which spews new carbon dioxide into the atmosphere) such a process on Mars is a "one way trip." For, without plate tectonics to recycle the "locked up" atmosphere, the carbon dioxide remains in the ground as buried rock (!), forever unavailable to the planet's atmosphere.

It is for these two, independent reasons, that I favor the idea that Mars truly "died" several billion years ago. Whatever catastrophes Bill Beatty is seeing probably happened long, long ago — and are still preserved (along with the "old craters") in a frozen tomb.

Which leaves unanswered the question: how could living, breathing "humans" evolve. . . or even live. . . on a planet with its atmosphere locked up in the very ground they walked on?

We will try to answer this single most important environmental question relating to this Investigation in our next Entry.

For two very important reasons Hoagland leans toward the position that little has happened on the Martian surface for billions of years. Consequently, his hypothetical civilization would be vastly older than the human race on Earth. Lambert's next entry adds credence to Hoagland's assertion. Michael Carr of the United States Geological Survey is one of the foremost experts on Mars.

Once again, disasters are in vogue.

[124] Dolphin [Lambert] 28-Feb-84 2:16PM-PST

Bill Beatty and Michael Carr of USGS (author of "The Surface of Mars") met today. Carr believes a sudden disaster occured on Mars sometime in the first one quarter of the history of Mars. He has no explanation. He suspects a sudden pole shift. He concurs that the early history was governed by water-flow, probably a dense humid atmosphere and warm climate. Permafrost now occurs to 3 kilometers or more, which is extraordinary. Whatever caused the disaster was permanent, the climate being cold and the atmosphere thin ever since. Carr says they have just completed hardware development plans for the next Mars orbiter which will not have a camera but radar imaging system, gamma ray detector, etc., and lifetime about 2 years. Carr thinks we have enough pictures of Mars, though as conference participants we might want to disagree with this in a few instances. Carr's book is excellent if any of you have not seen it. Beatty

NASA frame 086A08 showing the Utopia region of Mars. Landmarks include the Runway, Hangars, and mesas similar to that occupied by the Face. This region is on the opposite side of Mars. The runway could have been formed by a fault.

and Carr are in basic agreement on most of the features I summarized from Beatty's notes to me yesterday.

Lambert Dolphin now returns to Cydonia. Brandenburg has forwarded a copy of 753A33. Overall, it is an impressive confirmation of the bilateral symmetry of the "Face." However, Dolphin concludes that this view of the face indicates a completely natural formation. Later in entry #164, Dolphin will realize and admit that he was looking at the wrong formation. For the time being, though, this mistake disrupts the communication. Here is another striking example of one of the limitations of computer conferencing at the present time, since the participants could not share the image at the same time as they can in a regular meeting.

It is also important to note the way in which this type of computer conferencing eventually compels the correction of mistakes. Errors of this type were also made by Hoagland, Brandenburg, and DiPietro. One of the main features of computer conferencing, from a behavioral standpoint, is that unlike face-to-face meetings the discussion of ideas does not follow a pecking order. Within a computer conference activity or electronic "room", there is a leveling of the participants. The immediacy and intensity of the communication makes it less threatening to point out or to admit errors. Since participants have to read what the others are saying and have to compose their responses in writing, the interpersonal dynamics of an electronic meeting are very different from conventional meetings.

Note 1066 *From Dolphin* [*Lambert*] *to Hoagland* [*Richard*] *29-Feb-84 12:55PM-PST*

Just talked at length with Brandenburg and also took 753A33 down the hall where Beatty and I looked at it under a microscope. John examined it only in haste and is not surprised at my analysis. The image is free of noise (salt and pepper) in the vicinity of the face except for one salt dot near the left eye and one pepper dot just below. The sun angle is 45 degrees from the east, thus the face casts a shadow on the left side. The shadow is a triangle, with the nose at the apex. The southern side shows no chin, instead looks like a pyramid face with the inverted V in the middle. There is a bit of shadow or dark stuff right at the base of the south side which gave John the impression of a "helmet" (I think). The sunlit right side appears very steep and the right eye is very low down in elevation on the north face. The salt noise burst makes it difficult to make out the left eye clearly, but it also seems low on the north face. There is nothing in this image at all that gives any hint that this is a carved face. Off to the left is our city, of course, but all those pyramidal features stretch out for miles to the southeast. They all seem to be part of the same geology. There are many angled mountains and pyramids to the south of the face. I am tempted to take some modelling clay and sculpt the face as it appears in 753A33, then light the model from the west at a low angle. I think this will show the face is entirely natural. I will blow up the entire image and also try to enlarge the head by 10X if possible. John's calculations indicate that the eye sockets are about one pixel-size so this must be taken into account. John very correctly points out that this image mainly shows us the head is highly symmetrical. John also points out that SPIT processing on this ought to be helpful.

NASA frame 086A07 showing another region in Utopia, not far from the Runway. Features here include Envelope Island and the Mining Operation.

CHAPTER EIGHT

Martian Anthropology

Now that some of the basic geological and climatological information was in hand, I felt comfortable enough to generate some thought experiments. These entries were composed off-line and then downloaded on to the computer conference. As a result they are a lot clearer than my earlier entry.

[137] *Pozos* [*Randy*] *29-Feb-84 11:21PM-PST*

Some Notes on Anthropology and Mars

In entry # 30 on 3 Jan 84 I outlined some of the implications of this research for re-working the social sciences and the humanities. I concluded by trying to frame some of the questions raised by this research even if the landforms are found to be the product of completely natural forces. Now, let's take a look at the other possibility that these landforms are the artifacts of intelligence.

A Civilization on Mars

1. The first question that arises is whether the makers of the artifacts of Cydonia were human or were members of another species.

From our own experience with *Homo sapiens,* cultural artifacts reveal representations of the species. Other species are sometimes depicted in a variety of media but are usually done so to reflect human values or human social organization. Totems are good example of this.

The creation of the Face would not be an unusual thing for humans to do. However, this does not mean that it was made by humans. If it was indeed constructed, it may have been made by another species for a variety of purposes.

2. The age of the artifacts is the next crucial element.

Hoagland's use of celestial alignments in attempting to establish some concept of age for the artifacts is a very reasonable approach in the absence of other data. One half million years ago vastly precedes the development of *Homo sapiens* as we currently understand it. In fact our "ancient" monument-building civilizations don't go back more than 5,000 years.

The species which created the Martian civilization was either indigenous to the planet or immigrant. The development of an indigenous species, if it proceded in a manner similar to our conjectures about the development of *Homo sapiens*, would have taken at least 4 million years, assuming the previous development of suitable predecessor species.

The major drawback for the assertion that the Martians were indigenous is the absence of a suitable planetary environment over a major period of time to have made this development possible.

Even if we assume that an intelligent life form developed in the early history of the planet at a time when the sun was dimmer and Mars had a substantial atmosphere, such a development would precede analogous developments on earth by millions of years. Explaining such an accelerated development with our current knowledge is difficult.

3. If, however, these artifacts are indeed the remains of an exceedingly ancient civilization there could be some startling implications:

A) Humanity is ancient and has at least one other manifestation in the universe.

B) If this ancient civilization was organized around some state religious cult of its male ruler, it would indicate that the perfectibility of *Homo sapiens* is decidedly limited.

C) If this ancient society was destroyed or destroyed itself, the prognosis for the human condition is not optimistic.

D) If the ancient civilization was made by another intelligent species other implications occur:

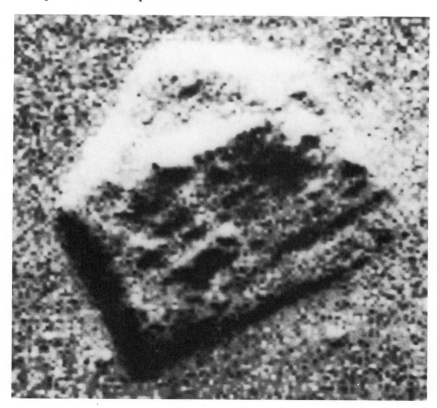

Blowup of NASA frame 086A07 showing Envelope Island in Utopia. This formation may be caused by debris heaps.

1. "Humanity" and "civilization" occur in other species which are not physically identical to *Homo sapiens* in ways and patterns very similar to those experienced by *Homo sapiens.*

2. A refocused exploration of the solar system could reveal more artifacts from other intelligent species.

Breck's excerpts from Bronowski's "The Ascent of Man" are a useful compendium representing a consensus in the field of anthropology. However, as with any textbook presentation, there are many controversies which are not presented.

The recent rise in catastrophism's respectability and apparent flexibility of genetic crossovers — as documented in the work of Barbara McClintock, who won the Nobel prize this year — indicate that the physical and biological development of the solar system does not always follow neat curves.

Physical anthropology and primate studies are areas rife with controversy. The fossil record is far from complete. The missing link has not been found and may never be. The concept of some common ancestor linking *Homo sapiens* with other species which are morphologically similar is reasonable but it also underscores the absence of that link and lays bare an assumption which could be upset by encountering extraterrestrial intelligence.

In social cultural anthropology, the emergence of stratified hierarchical patriarchal societies appears to be recent. Cultural achievements do not appear in neat ascending curves — Bronowski to the contrary — and many of the values which we currently cherish and use to define "civilization" have been existence less than 3,000 years.

From the standpoint of our current knowledge, the independent development of a species similar or identical to *Homo sapiens* on Mars is very unlikely. However, information from Mars and other parts of the solar system could recast our understanding of the Earth's record. Perhaps "consciousness", "intelligence", and "civili-

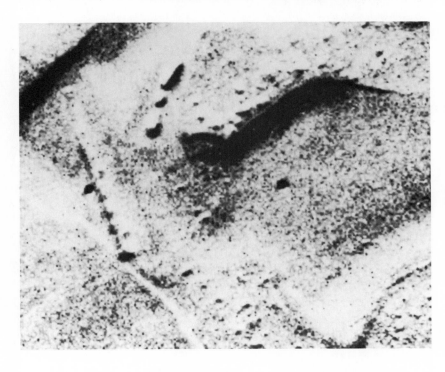

Blowup of NASA frame 086A08 showing the Runway in Utopia.

zation" require some neurological basis in material organisms as a necessary condition. However, human morphology or neural structure is not the sufficient condition.

Interestingly, we do not call our species *Homo cognoscens* (Man the knowing), *Homo cogitans* (Man the thinker), *Homo putans* (Man the speculator), but Man the wise/tasting. We may find on Mars information which will make us all the wiser.

Before responding to my entry, Lambert Dolphin completes his response to John Brandenburg and adds some additional observations about the "Face" and the "City." From the additional height of this image, several features appear to be remarkably different. At this point Dolphin is still refering to the *wrong* formation! Later in entry #164, when Dolphin realizes the mistake, he will look back on this analysis in dismay.

[138] *Dolphin* [*Lambert*] *1-Mar-84 6:00AM-PST*

In response to a note from John, yes the right eye is visible on 753A33 and I do see a distinct inverted V where the chin should be, like an erosion gully. Also, John, the base of the hill does suggest a "motorcycle helmet" instead of the hairline on the top and sides of the other frames. In spite of the low resolution I should point out that the complex features of the fort off to the east are all clearly visible on 753A33. Under the chin there is a pronounced shadow. Since the "chin" looks more like a pyramid face the shadow looks like a horizontal cut. Also there is a pronounced hill on the right top (A bump on the head). The nose is very high compared to the eye sockets so if the hill were a face it now has a ghostlike appearance, more like a skull than a face. However the hill on 753A33 does not really look like a face at all, more like a starfish or four-sided, badly eroded pyramid hill with two craters low down on the north face. The face hill is really very similar to many other of the bumps in Cydonia on this frame. I will probably have prints in the mail to everyone early next week.

In his response to my entry, Lambert Dolphin once again opens up more controversial areas of speculation. Dolphin alludes to an impressive visitor and brings in the myth of Quetzalcoatl from Mexico and Central America. Essentially, Quetzalcoatl was a prince and religious leader with a fair complexion who left toward the east. He promised that he would return from that direction on a strange beast wearing his characteristic turquoise. Hernan Cortez came to conquer Mexico about the time that the priests had predicted the return of Quetzalcoatl. He came from the east on a horse, an animal unknown to the natives, wearing turquoise.

[139] *Dolphin* [*Lambert*] *1-Mar-84 6:22AM-PST*

Thanks, Randy for your fine entry. It occurs to me that non-human life forms on Mars might carve a human face if they had a human visitor and were sufficiently impressed (the Quetzalcoatl white man scenario in Central America). We put statues of famous people in our parks to memorialize them and of course there is Mt. Rushmore. The Sphinx is believed to be Pharaoh Chephren, a son of Cheops (c. 2600 B.C.) and there are giant statues in Egypt such as the

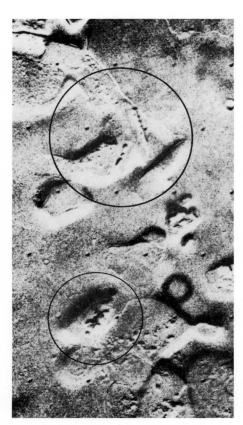

NASA frame 086A08 showing the Runway with Water Tank and more Honeycomb formations.

colossi of Memnon at Thebes and some big statues of Ramses II, the supreme egotist who ruled about 70 years. In Egypt the Pharoah identified himself with the son god Ra and was thus deified. Such monuments on earth point out horizontally for the benefit of fans or worshippers. A face on Mars pointing upwards would not likely evoke much reverence from the natives on Mars and seems to me to be more likely a signal to earth or elsewhere, perhaps even a call for help from a dying civilization in need of rescue. I am wondering what you think was the purpose of those huge Incan images that make sense only from the air? In the event the face were to go away as a real artifact, we still have numerous pyramids which look suspicious and they suggest a knowledge of mathematics and astronomy which non-human life forms might develop as readily as man. Incidentally, regarding the pyramids of Egypt, the report of Herodotus is often quoted to the effect that the pyramid of Cheops took 100,000 men 30 years to build (ten years for the ramps). So as Kurt Mendelssohn [*The Riddle of the Pyramids*] points out, the pharaoh had to mobilize most of the work force and a large fraction of the GNP to build a pyramid. This means laying a 2-3 ton block every 30 seconds or so if the work was done only during the 3 month flood of the Nile, and for 20 years. Our Martian pyramids are much larger, though gravity is less. Why build things so big if they were strictly utilitarian? If monumental then a very extensive civilization must have been there also so as to afford such luxurious monuments. A civilization on Mars might not have destroyed itself but been the victims of a terrible accident, such as a pole shift or climate change with loss of atmosphere, forcing them to take desperate action to survive a little longer. Alvarez even suggests that the dinosaurs didn't necessarily die out in one day but over a period of years. However the 50,000 frozen mammoths in Siberia were deep-frozen fast as stomach contents are intact and flesh still edible. I am especially glad you called attention to both catastrophism and to unanswered questions in the fossil record. Because civilization has been in existence less

NASA frame 086A10 showing Brandenburg's ''other faces'' in the Utopia region. The upper face is ''Comedy''; the lower is ''Tragedy.''

75

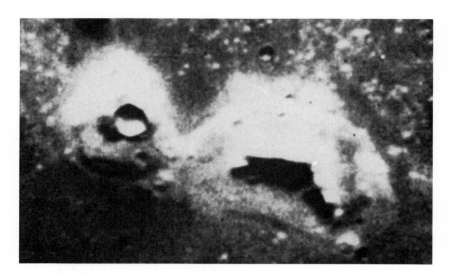

Closeup of NASA frame 072A04 showing the Hangar in the Gate Pyramid area. [SRI]

than 10,000 years, this has always caused me to view time scales of billions of years with a jaundiced eye, since I believe the universe was made for man and makes little sense standing empty for most of its lifetime.

Once again, Lambert Dolphin reveals his theological leanings. Candidly, he states his belief in the traditional Judaeo-Christian notion of the creation of the universe for humanity. Traditional Christian theology has taught that the creation of the universe was a natural extension or overflow of the creative love of God. Traditionally, this has been expressed in the saying, ''Good is diffusive of itself.'' This belief is also gaining support from physicists studying the origin of the cosmos due to the very low probability of all of the events required for the structure of matter, energy and life. Physi-

Blowup of NASA frame 072A04 showing teardrop-shaped Gate Pyramid Island in center of image. The Hangar appears at the bottom of the island. [SRI]

cists Paul Davies, John Wheeler, and Stephen Hawking all allude to the possibility that the universe is the way it is because we are here to conceive of it and perceive it.

On a more pragmatic note, Dolphin opens up the question of the Martian economics by reflecting on the great cost of the Egyptian pyramids.

[140] Dolphin [Lambert] 1-Mar-84 7:59AM-PST

I have an interesting article on the approximate cost today ($1.1 billion) to replicate the Great Pyramid, which measures 700 feet on the base, is 450 feet high and contains about 2½ millions blocks. I have sent several of you copies but if I missed anyone who would like a xerox, drop a note.

By now the curious weaving of entry and response in the conference should be readily apparent to you. Computer conferencing has a curious "loop de loop" thought structure which takes a lot of getting used to. In response to Lambert Dolphin's previous questions, I now loop back to upward-facing formations on earth, some of which were the bases for Von Daniken's speculations contained in *Chariots of the Gods*.

[142] Pozos [Randy] 2-Mar-84 8:16PM-PST

In response to Lambert's question about the upward orientation of the face and the similar orientation of structures in Peru, here are some thoughts.

1. There may be a simple explanation. Namely, that the geological formations suggested the image of a face. In addition, the alignments were convenient and the mesa was sculpted to become a suitable and

Portion of full-frame NASA image 072A03 showing entire Gate Pyramid area.

useful calendar and monument with many layers of significance for those who built it.

2. In the Mojave desert, there are long serpentine artifacts that extend for several miles and coincide with the transformation of the area from lush grasslands with plenty of game to the present desert. The current consensus among archaeologists is that these structures were ritual attempts to bring about the return of the previous favorable conditions.

3. Similar climactic evidence is lacking in Peru, although most mythopoetic systems have earth-based or cthonic deities and sky-based or celestial/solar deities which overcame the cthonic or earth deities. Earth-based deities are generally feminine and are relegated to the dark underworld, the night, or the moon. Celestial/solar deities are generally masculine and are relegated to the sky or high mountains, the day, or the sun. Upward and downward are usually integral facets of most belief systems and are related to the four directions and dimensions as well as to the fifth dimension and direction, the "inner direction," which represents the forces of anti-matter/time. These "primitive" notions have recently gained a certain respectability due to recent developments in theoretical physics. Consequently, the upward orientation of the face has many rich interpretive possibilities from the stand point of *Homo sapiens.*

4. Cultic monument-building among *Homo sapiens* to bring about the return of favorable conditions is still prevalent in many religious and secular practices such as devotional shrines or tourist attractions. Somehow visiting special places provides an opportunity to share the charisma of historic persons or events. In fact, cultic monuments generally provide an opportunity to transcent space-time to communicate with the dead or to deal with past events which have gone into the fifth dimension.

5. The recent completion and dedication of a monument to the fallen in the Viet Nam war is a striking example of the personal, social and political dimension of this type of structured human behavior.

Overview of teardrop-shaped Gate Island.

Even if we are loathe to interpret many of our current urban behaviors in such a "non-scientific" light, the cargo cults of the Pacific provide a more distant example (from our perspective) of this feature of the social activity of *Homo sapiens*. People participating in the cargo cults construct landing strips and control towers replete with windsocks in order to receive gifts which their ancestors are sending them. The planes never land anymore (since World War II) because of spells, but will return once the spells have been broken.

6. From the standpoint of *Homo sapiens* an upward facing monument could very well be an attempt to communicate or to attract attention. However, if we are not dealing with *Homo sapiens,* it is hard to extrapolate, unless what we call "humanity" is a characteristic of reflexively conscious life forms.

7. Lambert Dolphin's comment on the likelihood of other intelligent species existing in the universe, since there "should" be someone there to perceive it and make use of it, is a poetic argument of great beauty. This approach hearkens back to ancient and medieval philosophers who argued for the existence of higher orders of being and an objective meaning for the universe due to its sheer order, magnitude, and wonder. Far from being "unscientific," this type of argument is based on the very nature of scientific questioning generated by the experience of wonder and awe and should remind us all of the fundamental well-spring of science from which the ethics of our research also rise. Science — whether physical or social — emerges from awe, the experience of the good and the beautiful, and from the love of wisdom [*philosophiae*] which is its origin and terminus.

Brandenburg continues the discussion of image 753A33 in his next entry. He describes his impressions of the face as seen in this image and then turns to a discussion of the scientific process and importance of the analysis of these data. He notes that even if the objects are ultimately proven to be of natural origin, the investigation will have been a success scientifically because of the process employed.

[143] *Brandenburg [John] 2-Mar-84 9:16PM-PST*

THANK YOU RANDY, I HAVE NEVER HEARD A REASONABLE EXPLANATION OF THE NAZCA LINES AND OTHER UPWARD VISIBLE MONUMENTS BEFORE THIS. I AM VERY INTERESTED IN THE "MIND SET" OF BOTH ANCIENT AND MODERN MAN AND THEIR UNREALIZED SIMILARITIES. NEW MORNING PHOTO OF FACE: VINCE INFORMS ME THAT HE HAS FOUND IN HIS OLD NOTES THAT THE HEAD IS IMAGED ON 735A34. THIS WILL BE AN ADDITIONAL MORNING SHOT. GIVEN THE POOR RESOLUTION OF 753A33 AND THE PRESENCE OF STATIC, AN ADDITIONAL PHOTO WILL GREATLY LESSEN THE CHANCES OF MISINTERPRETATION OF THE IMAGES. I WILL ORDER THIS ASAP.

COMMENT ON THE MORNING PHOTOS: AS THE ONLY OTHER CONFEREE WHO HAS AT THIS POINT SEEN 753A33 AND EXAMINED IT CAREFULLY, I FEEL OBLIGATED TO

NASA frame 072A02 showing area adjacent to Gate Pyramid. Here the pertinent feature is Ayers Rock on Mars. This formation is 8.8 Km long by 2.9 Km wide.

SAY SOMETHING. MY EXAMINATIONS WERE WITH A MAGNIFYING GLASS, NOT AS GOOD AS LAMBERT'S MICROSCOPE, BUT GOOD ENOUGH TO SEE DETAILS DOWN TO A PIXEL SIZE. MY PRINT WAS SENT TO LAMBERT SO YOU COULD ALL SEE IT AS SOON AS POSSIBLE.

MY OWN IMPRESSIONS WERE THAT THE PHOTO CONFIRMED THE OVER-ALL SYMMETRY OF THE HEAD, INCLUDING ITS SUPPORTING STRUCTURE. THE FACE WAS ALSO SHOWN TO BE FRAMED COMPLETELY AROUND, AS IF WEARING A HELMET WITH A HEAVY, ARMORED CHIN STRAP. (I HAVE THOUGHT THIS WAS A HELMET ALL ALONG.) THESE FEATURES ARE REVEALED BECAUSE THE LIGHT IS FROM BELOW AND FROM THE RIGHT. SINCE WE HAVE NEVER HAD THESE THINGS ILLUMINATED I CONSIDER THIS IMAGE AND THE ONE FROM 753A34 WHICH WILL GET TO YOU ALL SOON, HOPEFULLY, TO BE GOLD MINES OF INFORMATION. HOWEVER, SINCE THE PIXEL SIZE IS SO LARGE (5X 35A72) AND THE LIGHTING ANGLE IS HIGH (45 DEGREES), THE IMAGE MAY LOOK SOMEWHAT PECULIAR IN DETAILS NEAR THE PIXEL SIZE. SINCE WE HAVE ALREADY TWO IMAGES AT MUCH HIGHER RESOLUTION, I WOULD CONSIDER OUR IMPRESSIONS OF THEM TO BE FAR MORE RELIABLE.

IT IS OF COURSE POSSIBLE THAT CAREFUL ANALYSIS OF THESE TWO MORNING PHOTOS WILL SHOW THE OBJECT'S RESEMBLANCE TO A HUMAN FACE TO BE AN ILLUSION. THIS WOULD BE A SERIOUS BLOW TO THE HYPOTHESIS THAT IT IS FACT AN ARTIFICIAL CONSTRUCTION. THIS, DESPITE THE FACT SEVERAL OTHER OBJECTS IN THE AREA SEEM TO BE ARTIFICIAL CONSTRUCTIONS. THE HEAD, IN THE ABSENCE OF ANYTHING AS STARTLING IS IN FACT A CENTRAL PIECE, IF NOT THE CENTRAL

PIECE OF EVIDENCE FOR THE HYPOTHESIS THAT A CIVIL-
IZATION ONCE EXISTED ON MARS.

I MYSELF FIND THE PROSPECT THAT ANY NEW PHOTO
WILL SHOW THE HEAD TO BE ANYTHING MUCH DIFFER-
ENT THAN WE SEE IT NOW TO BE HIGHLY UNLIKELY.
753A33 SHOWS THE HEAD TO HAVE A HIGH DEGREE OF
SYMMETRY, IF IT HAD NOT SHOWN THIS I WOULD HAVE
IMMEDIATELY DOUBTED IT COULD BE ARTIFICIAL, HOW-
EVER IT INSTEAD CONFIRMS THE SYMMETRY SEEN IN
70A13. THIS PROVIDES MORE EVIDENCE THAT IT IS
ARTIFICIAL.

ALL TRULY INTERESTING SCIENTIFIC INVESTIGATIONS
ARE FULL OF DRAMA AND SUSPENSE AS DATA TRICKLES
IN (SLOWLY AND NEVER ENOUGH AS YOU'D LIKE) AND
HYPOTHESES AND THEORIES ARE EITHER DESTROYED
OR CONFIRMED. I HAVE EXERTED MYSELF TO FIND EV-
ERY USEFUL IMAGE OF THIS OBJECT FOR THIS CONFER-
ENCE SO THAT WE CAN REACH THE BEST CONCLUSIONS
POSSIBLE, IF IN FACT THIS AND THE OTHER OBJECTS OF
INTEREST CAN BE SHOWN TO BE MERELY GEOLOGIC
FEATURES I WILL CONSIDER THIS INVESTIGATION TO
HAVE BEEN COMPLETELY SUCCESSFUL. I WOULD OF
COURSE BE SOMEWHAT DISAPPOINTED, BUT WE WILL
ALL HAVE DONE OUR DUTY.

THE FACT IS HOWEVER, THAT WE (IN ALL LIKELIHOOD)
HAVE ALREADY FOUND THE ONLY USABLE IMAGES OF
THIS OBJECT. I FEEL THAT WHEN THESE IMAGES ARE ALL
CONSIDERED CAREFULLY BY THIS CONFERENCE, THEY
WILL ALLOW US TO MAKE A STRONG JUDGEMENT AS TO
THE PROBABLE NATURE OF THIS AND THE OTHER OB-
JECTS OF INTEREST IN CYDONIA.

COMMENT ON 70A13: I HAVE TODAY UNDERSTOOD
WHY THE EYESOCKET ON THE RIGHT LOOKS LOWER
THAN LEFT. ANSWER: THE SHADOWING IN THE RIGHT
"EYE" CAN BE UNDERSTOOD AS BEING DUE TO LIGHT
FROM THE HILL BESIDE THE HEAD ILLUMINATING THE
FACE FROM THE LOWER RIGHT. THIS LIGHT CAUSES
SHADOWS IN THE LOWER EYE SOCKET AND ILLUMINATES
THE REST. THE LOWER SHADOW EDGE IS THUS THE
BOUNDARY OF THE EYE SOCKET. THIS LOWER EDGE CO-
INCIDES VERY CLOSELY TO THE LOWER EDGE OF THE
LEFT EYE SOCKET. ALSO, THIS EXPLAINS THE CURIOUS
SHADOW SEEN IN FALSE COLOR IN FIG. 37 OF UNUSUAL
MARTIAN SURFACE FEATURES. THE CHEEKS SEEM TO IN
FACT LOOK "SKELETAL" WITH CAVITIES BELOW THE
EYES NEXT TO THE NOSE. THIS APPEARS AS AN EXTEN-
SION OF THE "EYE ORNAMENT" VISIBLE ON THE LEFT EYE
IN BOTH HI RES PHOTOS. THIS WOULD CONFIRM "GHOST
LIKE" FACE SEEN IN 753A33 SEEN BY LAMBERT. LAMBERT,
IS NOSE SHADOW ON LEFT "SPLIT" BY EYE SOCKET EDGE?

I'M CONSTRUCTING A CLAY MODEL.

WARP SPEED! JOHN.

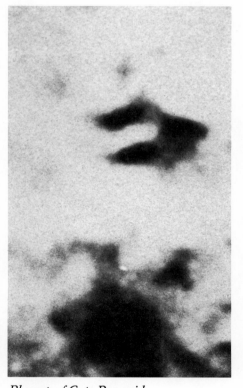

Blowup of Gate Pyramid.

Lambert Dolphin begins his next entry with a reference to the still wide-ly accepted notion from the European Renaissance that humans are the measure of all things. In religious circles this is still considered highly sus-pect but it is not rejected due to the Judaeo-Christian notion that humanity is the apex of earthly creation. Humans as the measure of all things was the critical link from which the secular science of the twentieth century was to grow. As we are about to enter the twenty-first century it is clear that we are beginning to define humanity more broadly than our own racial or eth-nic group. The next challenge in space exploration will be arriving at a star-system (or perhaps galactic) identity.

[144] *Dolphin* [*Lambert*] *3-Mar-84 4:58AM-PST*

1. A number of scientists in the past have commented that man lies midway on the scale between the very small and the very large. Thus the universe appears to center around us. An expecially good recent book is "On Size and Life" by Professors McMahon & Bonner of Harvard and Princeton in a Scientific American Library series.

2. If I did the arithmetic correctly the four-sided DiP&M pyramid on Mars has a volume 176 times that of the great pyramid of Cheops at Giza. Allowing for the lesser gravity on Mars, building such a structure does not seem to me to be unreasonable to the nth degree. The builders could have had machinery to help, whereas the ancient Egyptians did not have more than copper chisels, rollers, ropes and levers as far as I know.

3. John, as far as I can tell, the shadow of the nose is all dark. . . a triangle (in 753A33). Will indeed be anxious to see 753A34.

4. Randy, I, too, greatly appreciate your helpful and enlightening last entry.

Blowup of larger area of Gate Pyramid.

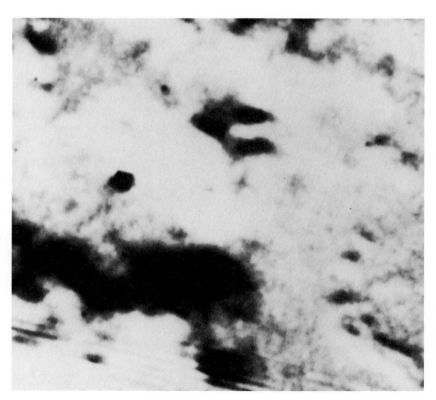

CHAPTER NINE

Ruins of Pyramids

[146] Hoagland [Richard] 3-Mar-84 2:18PM-PST

Several comments:

1. Randy's two entries regarding the anthropological implications in this Investigation — both extrinsic and intrinsic — are truly first rate. I will have some additional thoughts on the intrinsic implications as part of the continuation of my "mini-series": Some thoughts on Martian Climate, etc.

2. On Lambert's calculations regarding the D&M pyramid: It is FIVE-SIDED, not four-sided. Furthermore, if my impressions are accurate, the sides rise at about a 30 degree slope — indicating that the "summit" is about a mile (!) above the plateau on which it is "built." Regarding this last statement: It is my belief that it was NOT "built," but CARVED — out of a pre-existing mountain. I base this reasoning on several factors:

 A) The absence of any substantial "quarry" evident in the pictures, whereby the "stones" for its construction could have been excavated.

 B) The presence of substantial amounts of detritus around it. I interpret this (on the western and northern sides) as evidence of where they "threw the dirt."

 C) The substantial evidence that the southeastern side (matching the 1.6 mile northwestern side) has collapsed, with its material flowing out from the base in a fashion very similar to the famed Egyptian pyramid at Meidum. This catastrophic event was triggered, I believe, by whatever made the peculiar small crater near the northeast "buttress" still visible above the debris. If you carefully measure the length of the two southern "buttresses," you will find that the southeastern one is slightly shorter (by a couple hundred meters) than its counterpart — consistent with the southeastern cornice being partially buried in the debris from the collapse of the southeastern side.

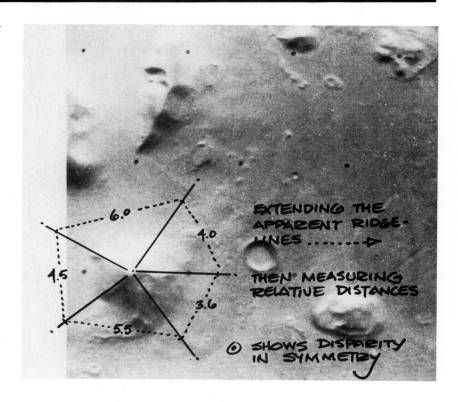

The shadows, debris fans, etc. all support this model. What is very interesting, however, is the WAY in which this side collapsed. It did not flow very far from the base of the pyramid, but remained piled around the base. In similar collapse features in other parts of Mars, however, the debris has flowed across level surfaces for tens of kilometers (!), and from a "fall" not equal to that encountered here. From the standard equations published on these other landslides, it is evident that something "lubricates" these extended landslides. The most likely candidate is water (as permafrost which locally melts, triggering the collapse.

The fact seems to be that we are NOT dealing with a similar phenomenon here. The flow pattern is too short for such internal lubrication. This, in turn, implies that the collapse of this sidewall was "different," both in its cause and in its final effect. For, though this site (at 41N) is loaded with permafrost features (see photograph 35A72 just to the west of the D&M pyramid), and a natural feature should retain a similar amount, the collapse of this "pyramid" argues that this object is somehow "different" from its surrounding objects. My perception is that this is because it was deliberately modified — from just another mountain into a five-sided pyramidal structure. My feeling is that it was hollowed out, vast numbers of chambers excavated inside, which weakened its support to the point that some external event triggered a substantial collapse of the southeastern side. The evidence supporting this can be found in the presence of the "square craters" which shadows reveal at the base of the northeastern "short-side" of this object. I feel that these are the joint collapse of many adjoining "rooms" in the interior, concurrent with the main catastrophe.

If this theory is true, it would explain why there is, indeed, "buttressing" at all the corners. It was both a way to reinforce an essentially hollow structure (as more and more rooms were excavated) as well as a "place to put the dirt" which would have the best effect!

Whether this was a "low technology" or a "high technology" project, I cannot say (yet). My feeling is that it was the first major pyramid in the area, after a lot of previous activity. I'll have some further thoughts on this in a future entry. But the so-called "bottom line," I believe, is that this (and the other, major pyramidal structures) were intended as living places — NOT tombs. Only the need to shelter a large number of people, and the absolute necessity to do so REGARDLESS OF THE COST, I feel, could account for the scale of these "monuments of Mars." And that, in turn, supports some other thoughts on the climate which was present when these things were built. More later.

[147] *Hoagland* [*Richard*] *3-Mar-84 10:20PM-PST*

John,

I've examined very carefully my copies (multiple contrast printings) of 70A13 — the "head." I find I cannot agree on your placement of the light source(s) which are creating the shadows. The reason is simple: shadow direction.

The detail I see is totally consistent with "grazing incidence lighting" from the sun, from the left. The shadow for the eye cavity certainly conforms to this. The shadows in the "hair" (or helmet) indicate a "stepped bench" lit from the left. I see the feature below the right eye which you tentatively identify as similar to the "tear" feature below the left eye. In fact, it looks to me like some optical aid or such.

More and more I agree with you, however (and with Jim Channon), that this figure is a military figure of some kind. And that has enormous implications, both in terms of the cultural significance for the society which had to create all that we are discovering, as well as for certain long-term myths associated with Mars. . .

Are we, in fact, looking at "the God of War," gentlemen?

[148] *Dolphin* [*Lambert*] *4-Mar-84 5:06AM-PST*

1. The great pyramid of Giza is 750 feet base and 450 feet altitude, corner angle 51 degree. To calculate the volume of the D&M pyramid, I used 1000 feet height. If you prefer a mile high, then the D&M pyramid has a volume 880 times that of Giza.

2. Mining operations are very expensive and slow. Conventional tunnelling proceeds at a few tens of feet per day and, if I remember right, sort of like $100 per foot. (I'll check this Monday). To carve out a solid rock pyramid would take many years on earth. Also removing a large mass of rock would probably make the mountain less likely to collapse under its own weight at least.

3. I see the D&M pyramid as four sided with one badly damaged face.

4. If a mountain were carved out inside, the easiest way is numerous entry adits all over the place, with resulting tailing piles which

would ruin the nice pyramid shape. If intended for emergency housing, why bother with pyramids? Holes dug in the ground will do fine in an emergency. The best reason I can think of for pyramids is religious or aesthetic, in which case one would think of the buildings as utilitarian and built in a time of prosperity and peace. The FACE could have been a signal to the gods as Randy says, but pyramids? Why complexes of pyramids in the city area?

5. Our highest available resolution is really still too coarse and I vote strongly for a high resolution camera asap on the strength of our many enticing artifacts that will surely be fully understood only when someone goes back for better pictures.

Inca City on a high plain near the South Pole of Mars. Taken by Mariner IX. [NASA]

[149] *Hoagland* [*Richard*] *4-Mar-84 9:52PM-PST*

Lambert,

My choice of a "mile high" for the D&M pyramid is not arbitrary. At the sun angle of 70A13 (about 30 degrees), the indicated slope of the shadow of the southeastern "buttress" is about 30 degrees — indicating that is the angle of incline to the southern side. That (in a pyramid measuring 1.6 miles on the long side) makes the "altitude" somewhat less than a mile.

The D&M pyramid is an impressive object!

I'll make an overlay sketch of the FIVE sides and send it to you, for transmission to the rest of the Team. We all have to be looking at and describing the same things, if we are to make headway in this.

I'll have some thoughts on "why excavated mountains" in my continuation of the Martian climate series. . .

[150] *Dolphin* [*Lambert*] *5-Mar-84 8:28AM-PST*

Addition to Bibliography on Mars:

Vivien Gornitz, Ed., "The Geology of Mars", Dowden, Hutchinson & Ross Inc., Stroudsburg PA., 1979. A collection of 36 papers, with comments on each by editor.

[152] *Dolphin* [*Lambert*] *5-Mar-84 10:37AM-PST*

Bill Beatty confirms that $100 per foot is a conservative cost of tunnelling for an 8 feet wide by 10 feet high tunnel. A Jumbo Drill with crew of three plus support can make 24 feet per day (three shifts). Tunnelling out a mountain is far harder than building a building from scratch. After 100 feet of tunnel one needs a fresh air supply such as a shaft to the surface. Any extensive interior chamber could be facilitated by multi-levels with waste drip shoring up except in hard rock like granite. Could take thousands of feet of timber or other supports. At New Almaden Mercury Mine south of San Jose there are 100 miles of tunnels and stopes (ore removal rooms) so large one could put the village church inside. Worked over a hundred year period by a work force of hundred men. Sorry for the noisy connection today.

Hoagland challenges Dolphin's reservation about the costs of the possible construction of such large structures by appealing to human history and provides an argument for an excavation hypothesis.

[153] *Hoagland* [*Richard*] *5-Mar-84 11:08AM-PST*

Re Lambert's entry 152:

Just because something's expensive, difficult, and makes no sense to us, does NOT mean historically that others obeyed our "common sense." The Great Wall of China and the Pyramids of Egypt (which Mendelssohn calls the "marvelous madness of the 4th Dynasty") are perfect examples.

If the D&M pyramid was erected, Lambert, where are the quarries? (In the frame to the west of 35A72 (35A74, I believe) my examination of the "wall" on the ejecta of the rampart crater northeast of the Face shows a definite excavation behind it. The volume of the excavation matches perfectly (!) the amount of material which was piled up to create the "wall." I don't find any similar feature within 30 miles of the D&M pyramid. I do find that its overall size matches other mountains in the vicinity, indicating that not much shaping may have been done to its exterior. As for excavating rooms in the interior, I point again to the suspicious square-shaped cavities (as outlined by the shadows) in the enlargements of this object. There is definite evidence of interior regularity, as I'll have Jim Channon demonstrate in a sketch. As to the reason: that's coming up.

CHAPTER TEN

Placing Bets

Dolphin begins a betting match about the probability that the landforms are constructed. This will begin the conclusion of the first phase of the conference by requiring the participants to summarize their positions and views.

[154] Dolphin [Lambert] 5-Mar-84 1:59PM-PST

Bill Beatty, our trusty senior geologist here at SRI, having looked at all the controversial images and read the references on Mars cited in the Bibliography, says he would guess there is a 5% probability that some of these artifacts are not natural. He is not especially conservative, quite ready to change his mind if new evidence is presented to him. For me, I would vote 25-30% probability on the basis of what I have seen so far. I would be interested to know if others would care to gamble for the sake of a preliminary survey?

[155] Hoagland [Richard] 5-Mar-84 2:18PM-PST

My perception of probability that objects we are seeing artificial: 80%
(But then I've had about 6 months to look this stuff over, and do a lot of measuring.)

John Brandenburg takes a broader perspective on the questions of whether the landforms are constructed by asking whether he can scientifically discount the possibility that they are constructed and not natural. His presentation is an example of fine orthodox scientific argument. Since Brandenburg's terminal could only enter capital letters into the InfoMedia computer his entries and notes were all received by the participants in upper case. In computer conferencing this can be disturbing since it makes it appear that the participant is constantly shouting. The effect in this case was to make the entry appear to be a draft communique from the conference — a rallying point for the participants.

IT IS APPROPRIATE, I FEEL, TO NOW LOOK FOR A CONSENSUS, IF ONE EXISTS, ON WHAT WE HAVE SEEN. DESPITE OUR EFFORTS TO FIND MORE HIGH RESOLUTION PICTURES OF THE HEAD WE HAVE FOUND NONE. MORE MAY EXIST, BUT ONLY AN EXHAUSIVE AND DETERMINED SEARCH WILL FIND THEM. SUCH A SEARCH WOULD HAVE TO BE DONE BY JPL OR ITS TAPE LIBRARY AND JPL WOULD NEED A COMPELLING REASON TO DO SO. WE HAVE FOUND THREE NEW LOW RESOLUTION PHOTOS OF THE OBJECT. WE HAVE TWO PRINTS OF THE THREE SHOTS. ONE IS AT MORNING AND SHOWS THE RIGHT SIDE OF THE FACE FULLY ILLUMINATED. THE OTHER PRINT SHOWS ALMOST THE SAME SUN ANGLE AS 70A13 AND ONLY CONFIRMS THE HEAD'S GROSS APPEARANCE DUE TO THE OVEREXPOSURE OF THE PRINT. THE MORNING SHOT SHOWS THE HEAD TO BE HIGHLY SYMMETRICAL AND FREE FROM ANY UNEXPECTED FEATURES. THE UNSEEN SHOT 753A34 AND A SUSPECTED FOURTH SHOT 673B57 OR 55 WILL ONLY CONFIRM THE VIEWS SEEN IN 753A33 AND 673B56 THEREFORE WE HAVE 35A72 AND 70A13 AT HIGH RESOLUTION. AND ONE VIEW 673A56 THAT CORROBORATES 70A13 AND ANOTHER 753A33 WHICH SHOWS A FUZZY SYMMETRICAL OBJECT LOOKING MUCH LIKE THE OBJECT SEEN IN 70A13 AND 35A72 SHOULD LOOK LIKE, IF IT WAS A CARVED HEAD, LOOKING UP AT THE SKY. A HIGH RESOLUTION MARINER 9 PHOTO OF THE AREA WAS FOUND, THE ONLY ONE APPARENTLY, BUT A DUST STORM OBSCURES THE SITE.

SO, GENTLEMEN, WE HAVE SOME NEW DATA, BUT MOSTLY WE HAVE GREG AND VINCE'S PHOTOS AND THEIR COMMENDABLE WORK ON THEM. THE NEW PHOTOS DO LITTLE BEYOND CONFIRM SYMMETRY AND SHOW AGAIN THAT THE OBJECT LOOKS LIKE A FACE IN A HELMET. I FIND MYSELF, DESPITE MY EFFORTS NOT TO, TO BE REACHING A FAIRLY FIRM, THOUGH PRELIMINARY CONCLUSION. I BASE THIS CONCLUSION ON SEVERAL FACTORS:

1. I KNOW THAT STRONG EVIDENCE POINTS TO MARS ONCE HAVING A DENSE ATMOSPHERE AND WARM CLIMATE: LONG WATER CARVED CHANNELS THAT COULD NOT HAVE BEEN MADE UNDER PRESENT CONDITIONS. ALSO EVIDENCE OF RAPID CRATER EROSION IN AN EARLIER EPOCH.

2. I (KNOW) THAT VAST QUANTITIES OF WATER EXIST FROZEN IN THE SOIL AND IN THE POLAR CAPS. THE WATER IN SOIL WOULD PROBABLY NOT BE THERE EXCEPT THAT LARGE QUANTITIES OF WATER STOOD ON THE SURFACE, THIS THE OPINION OF AN HYDROLOGIST, ROBERT BECHARD, WHO IS NOW STUDYING MARS FOR ME.

3. THE PRESENCE OF ENORMOUS QUANTITIES OF FEROUS IRON OXIDE, HEMATITE, ON THE MARTIAN SUR-

FACE GIVING IT ITS RED COLOR. FERROUS IRON, RATHER THAN BLACK FERRIC IRON, IS ONLY FOUND WHERE LARGE QUANTITIES OF FREE OXYGEN ARE PRESENT OR WERE PRESENT. THE PRESENCE OF FERROUS IRON IN PRECAMBRIAN ROCKS OF LATE EPOCHS AND ITS ABSENCE IN EARLIER ROCKS IS CONSIDERED BY GEOLOGISTS TO DATE THE CHANGE OF EARTH'S ATMOSPHERE FROM PRIMORDIAL TO OXYGEN CONTAINING, THE OXYGEN BEING PRODUCED BY ALGAE.

4. THE FACT THAT LIFE ON EARTH SEEMS TO HAVE BEGUN ALMOST IMMEDIATELY AFTER THE EARTH COOLED AND AS SOON AS MULTICELLULAR ANIMALS APPEARED, EVOLUTION ADVANCED AT A TELESCOPING RATE, PRODUCING MAN IN 500 MILLION YEARS.

BASED ON THESE CONSIDERATIONS, I CANNOT HONESTLY PRECLUDE THAT LIFE COULD NOT HAVE EXISTED ON MARS AT ONE TIME. THEREFORE, I DO NOT CONSIDER THE PRESENT HOSTILITY OF THE MARTIAN ENVIRONMENT WHEN I EVALUATE THE NATURE OF THE OBJECT. I KNOW TOO LITTLE OF MARS TO DISMISS ANY POSSIBILITY.

I THEN CONSIDER THE HEAD:

A) IT APPEARS COMPLETELY BISYMMETRIC

B) IT HAS TWO EYES A NOSE AND A MOUTH

C) IT APPEARS TO HAVE AN EYE IN ONE SOCKET AND ALSO: BY MY CAREFUL STUDY, TO HAVE CHEEK ORMENTS BELOW THE EYES.

D) IT IS PLEASING AESTHETICALLY, IT LOOKS A KING

E) OTHER OBJECTS OF NON-NATURAL APPEARANCE ARE FOUND IN THE IMMEDIATE AREA AND ELSEWHERE.

F) THIS SITE AND THE OTHERS APPEAR TO (BE) AREAS WHERE WATER WAS ONCE ABUNDANT: CYDONIA WAS THE PRIME VIKING LANDING SITE BUT LOOKED TOO ROCKY.

G) ALL OBJECTS RESEMBLING THIS OBJECT, FOUND ON EARTH ARE MAN-MADE. I HAVE NEVER SEEN ANY NATURAL FORMATION ANYWHERE OR HEARD OF ANY THAT APPEARED LIKE THIS, RESEMBLING A FACE TO THESE DEGREES (PROFILES DON'T COUNT AND WOULD FAIL MY CRITERIA: A,B,C)

H) I KNOW THAT SCI AMERICAN CARRIED AN ARTICLE SHOWING THAT EARTH WOULD LOOK LIKE MARS AFTER A PERIOD, IF ALL LIFE ON IT DIED.

I) THERE IS NOW EVIDENCE THAT CATASTOPHES OF SOLAR SYSTEM WIDE SCALE HAVE CAUSED MASS EXTINCTIONS ON EARTH.

BASED ON THESE CONSIDERATIONS, IT IS MY JUDGEMENT THAT THIS IS (AN) ARTIFICIAL OBJECT. I CONSIDER THIS HIGHLY LIKELY AFTER SEEING OUR BEST DATA. I KNOW FURTHERMORE THAT NO ACCESSIBLE DATA CAN BE FOUN(D)...

[157] *Brandenbury [John] 5-Mar-84 8:25PM-PST*

SORRY, HIT THE WRONG KEY.

WHAT I AM SAYING IS THAT (IF IT) WASN'T A CARVED HEAD, 753A33 WOULD HAVE IN ALL LIKELIHOOD HAVE SHOWN THIS CLEARLY. I FIND THE SIMPLEST EXPLANATION AT THIS POINT FOR THIS OBJECT'S APPEARANCE IS THAT IT IS ARTIFICIAL. THIS IS MY CONCLUSION AFTER SEEING THE DATA, AND I EXPECT BETTER ANALYSIS BY COMPUTER OF THE NEW PICTURES WILL STRENGHTHEN IT.

THAT IS MY OPINION: AND MY REASONS FOR HAVING IT. I BELIEVE ALL OF YOU WILL SOON BE FORMING YOUR OWN OPINIONS. IF YOU DO NOT SOON YOU (WILL) PROBABLY NOT FORM ONE, GIVEN OUR DATA ON THE HEAD WILL PROBABLY BE PRETTY MUCH AS IT STANDS NOW.

SINCERELY, JOHN.

In point #2 at the beginning of the entry Brandenburg plants a seed crystal which will substantially influence the direction of the conference and will anticipate the findings of NASA by a year and a half. His reference to water in the soil of Mars in large quantities is the beginning of what will become a discussion on the geochemistry of Mars in relation to the possibility of life.

In the next entry Dolphin agrees with Brandenburg but still returns to his reservations about the ''honeycomb'' structure.

[159] *Dolphin [Lambert] 6-Mar-84 4:47AM-PST*

Yes, I agree with John that additional new data is needed if we are to carry our studies further. Others can look over what exists and form their own conclusions, but to press forward, new inputs are needed to build a more convincing case if one can be built. The elusive honeycomb still needs a more definitive answer in my opinion (35A72).

Even as this section of the conference is winding down, Hoagland presses on for further work and better clarification through a comparison of the existing images. He also receives a copy of the enlargement of frame 753A33 which Dolphin has been examining and discovers the error in locating the Face which has been the cause of confusion in earlier entries.

[161] *Hoagland [Richard] 6-Mar-84 1:58PM-PST*

Comments on 753A33:

First of all, I sincerely hope we are all looking at the same object! I have two prints before me, the "raw" NASA frame (with data block, etc.) and an enlargement of a portion of that full frame. After a bit of searching, I located "the City" with respect to the bottom of the full frame (as determined by the data block). It is 1.5 inches above the bottom of the frame line, directly above the left-hand "Input histogram" data block. I can easily see the D&M pyramid, which is one and a fifth inches from the bottom of the frame, and also directly above the "Input" data histogram. The Face (which I found by lining up my familiar "sightlines" in the City) is one and a third inches above the bottom of the frame, exactly mid-way between the "Input" histogram and the "Process" histogram.

However. . .

On the enlargement Lambert had sent Special Delivery this AM, the Face is almost off the 8 x 10! It is exactly on the edge of the picture, 4 inches in from the left (if the large crater in the image is to the bottom right). Thus, another few tenths of an inch, and this blow-up would have missed the Face completely — leading me to suspect, Lambert, that you did not correctly identify it in the full frame, or that the photo lab did not follow your instructions.

[162] *Hoagland* [*Richard*] *6-Mar-84 2:16PM-PST*

Comments on 753A33 (cont'd):

My reason for assuming you made an error in identification is very simple: the object you described, "looking pyramidal, with one speck of "salt" and a speck of "pepper" matches exactly John's PALACE (which is due south of the Face), but NOT the Face in the enlargement I'm looking at. It seems to have two eyes, "low on the north side" as you described, with the speck of "pepper" south of the left "eye." But the object is NOT, I repeat, NOT the Face!

The Face (which is exactly one inch north of "the Palace" in this enlargement) appears very symmetrical, with an orientation at right angles to the City grouping off to the left. It has a definite centerline, which divided it EXACTLY in half, and two shadows — one stemming from the raised centerline (the "nose"), which falls over the "cheek" (as it would be in 35A72 or 70A13) but stops at the edge of the "platform" (or helmet, per Jim Channon or John Brandenburg). Below this, still to the left, a second shadow seems to originate from a combination of the base "platform" (or helmet) and the "eyebrow" projection above the left "eye."

It is not possible to see the northeast corner of the feature, as the edge of the photograph cuts off everything from the centerline to northeast of the right "eye" (as determined in 70A13). I can see no obvious right "eye" in this frame enlargement, although there are some albedo markings which are in the correct place. The left "eye" seems confirmed at this morning sun angle by the bright "rim" of the "eyebrow" ridge.

There is a really fascinating overall resemblance to a five-sided pyramidal shape, which intriguingly conforms to the five planes of the human face (an idea that is not original with me, but was observed by one Fred Lehrman, an associate of Marilyn Ferguson's, in L.A., a couple of weeks ago). Lehrman may have really hit on something. If so, it could explain the other five-sided pyramidal objects in this complex, as more and more stylized versions of "the Face!" But that is digressing.

I am equally intrigued at the subtle changes other "structures" in this area undergo at the higher lighting and completely different direction from which it comes. "The Fort" seems quite changed, from what we see in both 35A72 and 70A11.

In summation:

At this new lighting angle, frame 753A33 does not resemble a Face, so much as a five-sided pentagon (!), with a raised inner centerline. The new image does, however, confirm the symmetry of this strange object, allowing one to "fit" the unfamiliar view into the much higher resolution data we have at the opposite lighting. One thing is certain: we need higher resolution data!

[163] Hoagland [Richard] 6-Mar-84 2:59PM-PST

Comments on 753A33 (cont'd):

The geometry of the five-sided D&M pyramid is beautifully confirmed by this new data. Even the subtle "sweeping curve" of the left face (as first seen in 70A11 enlargements) is verified by the shadow caused by the morning light! The destruction to the eastern/southeastern faces is beautifully highlighted, and should have much detail when properly printed and SPIT processed. There is much to see in the overexposed portions of this image, which SPIT and a proper "stretch" should reveal. I am hopeful that such techniques will allow a better match between the features we see in the Face in this frame, and those in the other frames we possess at higher resolution. I, however, urge the "similar resolution comparison" brought up by Vince, in mentioning my comments in Entry 135. I think we can all learn a lot about image processing and interpretation by comparing different views at similar and opposite sun angles.

[164] Dolphin [Lambert] 6-Mar-84 3:04PM-PST

Dick is correct; I had mis-identified and described the wrong feature on 753A33. All my earlier remarks (gad) appear to apply to what John calls the palace. John, do check this out on your copy of 753A33 and confirm. The feature Dick is now describing doesn't look like our old familiar face but at least we have now agreed on the hill we are talking about! The left eye looks low in elevation, and the center line of the nose I see as a long ridge. The hill is quite symmetrical. I agree with Dick the change in lighting angle changes everything in the area!!

[165] Dolphin [Lambert] 6-Mar-84 3:08PM-PST

Yes, Dick is correct the D&M pyramid is a beautiful object in frame 753A33. I am very eager to see this frame processed and cleaned up ASAP.

At Dolphin's invitation Ren Breck, the conference sponsor, reflects on the work to date in the context of his interview with Nobel laureate Linus Pauling. The use of all capital letters in the entry, in this case, was done for emphasis, almost as if Breck were trying to be heard in a noisy conference hall by using a public address system.

[168] Breck [Ren] 6-Mar-84 11:05PM-PST

BEING PROMPTED BY LAMBERT, YOUR HUMBLE ORGANIZER IS PRIVILEGED TO SAY ONLY FOR THE MOMENT THAT YOU ARE ALL LIVING UP TO THE EXPECTATIONS WE HAD SET FOR THIS CONFERENCE. . . SCIENCE WHEN IT IS PURE REFELCTS THE HIGHEST ASPIRATIONS OF MANKIND, I THINK IT IS TRUE THAT WE ARE THE EARTH'S OWN EYES LOOKING BACK UPON HERSELF. . . WE ARE CREATION'S OWN CONSCIOUSNESS, THE CHILDREN OF

OUR HISTORY, THE FATHERS OF OUR FUTURE. . . TODAY I SPENT THREE HOURS WITH DR. PAULING AND INTER-VIEWED HIM ON HIS THOUGHTS ON MEDICAL PRACTICES, TREATMENT MODALITIES, AND THE FUTURE. HIS EIGHTY-THREE YEARS HAVE BEEN SPENT IN SEARCHING FOR TRUTH. . . HIS CALM ASSURANCE, AGAINST ALL RA-TIONAL ODDS, COMES FROM THAT SPECIAL ELEMENT OF ALL SCIENCE. . . THE COURAGE TO CHALLENGE ASSUMP-TIONS, THE HUMILITY TO LISTEN TO OTHER OPINIONS AND POINTS OF VIEW. HIS SPECIAL RADIANCE PERME-ATED THE HEARING ROOM AS HE SPOKE SOFTLY OF THE HUMAN RIGHT TO HEALTH, AND FREEDOM OF CHOICE IN TREATMENT. DR. PAULING'S RESEARCH INTO THE EF-FECTS OF VITAMIN C I THINK RIVALS THE WORK OF THE MEMBERS OF THIS CONFERENCE. . . AGAINST ALL TRA-DITION HE GENTLY CHIDES HIS PEERS TO TAKE A SEC-OND LOOK OVER THE TERRAIN OF THOUGHT. PERHAPS IT WAS HIS SPARKLING EYES THAT REMINDED ME OF THE FACE ON MARS LOOKING BACK UPON OUR TINY PLAN-ET. . . WONDERING WHEN AND IF WE WILL HAVE THE EYES TO SEE, AND THE COURAGE TO BE ALL WE CAN BE-COME.

The Level of Resolution

Hoagland addresses a fundamental concern raised earlier in the confer-ence by Gene Cordell and Lambert Dolphin. Many of the features are at the level of resolution of the images. How does this limitation affect the prelim-inary findings of the research? One of the main results was to create a great sense of caution. Dolphin's analysis of the wrong formation which was finally corrected in entry #164 only heightened this sense of caution.

[169] Hoagland [Richard] 7-Mar-84 12:17AM-PST

Comments on 753A33 (cont'd)
The limitations of using low resolution images to confirm details perceived at high resolution are excellently illustrated by 753A33. The D&M pyramid being a prime example. With the exception of the central (north) "buttress," (which is only discernable because we know exactly what to look for) the presence of these telltale (and currently inexplicable) features is totally missing on the 753A33 ver-ion of the D&M pyramid. Its key "difference" from the surrounding pyramid-shaped mountains, then, would be forever unknown — if we were limited to resolution of this quality.

By the same token, the absence of facial features on the Face in 753A33 says to me that that level of detail is simply not within the boundaries of the resolution in this frame.

The comparison with the companion shot (at similar resolution) should confirm this. Having said this, however, one can remark on the central features of these objects which ARE confirmed:

1) The overall geometry (five sides) of the D&M pyramid, and the mysterious destruction of two of those sides (Northeast/southwest).

2) The similar five-sided nature of the central pyramid in the City.

Gene Cordell Offers a Dissenting View

Facing the Future One afternoon in the summer of 1983, Ren Breck invited me over to his house for a dinner party. The subject of this dinner party was the "face" on Mars and some additional evidence for a "lost civilization". Ren asked that I come, examine the pictures, and give my opinion. I recalled the press reports from several years back, remembering that NASA had dismissed the "face" as an anomaly of the Martian surface enhanced by shadow and light. But new evidence? My curiosity was aroused.

I arrived at the party not knowing what to expect. Having spent four years examining Landsat satellite images, I was looking forward to seeing the images of Mars. After being introduced around, I inquired about the evidence for this "lost civilization". I was given several photographs with circles and arrows drawn on them.

Examination of the Pictures These photographs showed the "face" and another area which was being called the "city". The face was simple to analyze for it is there. It could be a natural formation of the Martian landscape or it could be something else. We can speculate on the face's origin, but the simplest explanation, the one requiring the least imagination, is that it is a natural formation.

But the area referred to as the "city", this was the principle evidence for a "lost civilization". It was, therefore, to the "city" that I turned my attention.

This "city was an area on the photograph whose principle feature was a honeycomb or crosshatched pattern. On close examination, I saw that the honeycomb pattern exactly coincided with the scan lines of the image and was of an equal resolution as the picture elements (pixels). My suspicions were immediately aroused.

To see why I was suspicious, it is first necessary to understand how the pictures were taken in the first place. Spacecraft are not able to take pictures in the normal sense because we would never be able to retrieve the film. Rather they take electronic pictures which are similar to the pictures on the television screen or in newspapers. If you examine a newspaper picture closely, you see that it is composed of tiny dots. Viewed from a distance and with a sufficient number of dots, our brains put the dots together into an image.

The spacecraft records these dots one at a time with a sensor mounted on a mechanical arm that swings left to right as it moves forward over the planetary surface. These dots are transmitted as radio waves back to Earth where they are composed into pictures. In the composed image, each series of dots, corresponding to the full swing of the mechanical arm, is called a scan line. If there is a malfunction in the camera, this will usually appear as dropped or blank scan lines. Similarly, if an artifact is introduced into an image, either through bad transmissions or through computer processing, this artifact will usually correspond to the scan orientation of the image.

Now, was it mere coincidence that the spacecraft's camera was perfectly aligned with the honeycomb pattern? Not likely. Not only did the honeycomb align perfectly with the scan lines, but each square in the honeycomb pattern was exactly the size of a dot in the image. These two pieces of information, taken together, mean that we are dealing with an artifact of the imaging technique. This information alone would suffice to identify the honeycomb as an artifact of the computer processing.

But to make the case conclusive, I needed to know more about how the image had been processed. Specifically, how many times and in how many ways had the image been modified. The image originated at NASA where some computer enhancements had been performed. Then it had been processed by DiPietro and Molanaar with another computer technique called Starburst Pixel Interleaving Technique (SPIT). From there they were transferred to photographic paper and enlarged. I was given a pamphlet on the SPIT processing technique which I sat down and read.

I learned that the SPIT process produced 9 dots for every pixel in the original image. The the "combs" or squares in the honeycomb pattern were not pixels from the original image as I had thought, but constructed dots created by the computer through the SPIT process. Additionally, the 9 new dots created by this process, taken together, produce a Starburst pattern out of an original pixel. Put several Starbursts together and you have a honeycomb. Thus you would expect a honeycomb-type pattern as a side effect of this particular computer processing technique.

This additional information was conclusive; the honeycomb pattern is an artifact created by the computer processing and does not exist on Mars. And, without the honeycomb, the case for the idea of a "city" on Mars becomes very improbable.

Some weeks after the party, I received a magnetic tape with the relevent raw, unprocessed Martian images. I took them over to my former employer, International Imaging Systems, and, with the assistance of a friend and current employee, put them up on their digital image processing equipment. We examined the five images on the tape. We looked at the area called the "city" and saw nothing of the honeycomb pattern. It did not exist on the raw image. We continued examining each frame, noticing the various Martian landforms, and the unique geology of this alien planet. There were many plateaus or mesas similar to the one on which the face appeared. Additionally, there were many "face" like structures on these mesas. With a little imagination we were able to see many faces. It was fun, but we had to conclude that the "face" was nothing but an interesting landform.

I passed this information along to the participants in the conference and as may be judged by the conference proceedings, it had little effect.

The Case for an Intelligent Face The evidence for the face having been created by some intelligent race depends heavily on the honeycomb. When I objected that the honeycomb was nothing more than an artifact of the image processing technique, I was for the most part, ignored. When DiPietro confirmed my assessment, stating that the honeycomb was indeed a manifestation of the computer processing and not real, these people had to regroup and find another way to establish the "city's" existence.

Ultimately, there is very little difference between the teleological thinking evidenced in this conference and the teleological thinking promulgated by the "creation scientists". The "creation Scientists" start from the premise that the Bible is absolutely true and then set about to support the Biblical story while casting doubt upon the accepted scientific theories. This conference started with the premise that the "face" was the creation of an alien intelligence, and was thereafter directed toward this very conclusion.

Randy Pozos, in his analysis of the different types of thinking that went into the conference, hit the nail on the head when he states that it depends on your epistemology and metaphysics as to what side of the argument you will be on. Exactly the same thing can be said of "creation science". If you are a born again, Bible as Truth, Protestant fundamentalist, then "creation science" makes perfect sense. It is not science, but from one point of view it makes sense.

This is not to say that there wasn't some real scientific inquiry involved in the conference, for there was. We learned through the course of this conference that Mars once had an atmosphere, oceans of water, and the basic ingredients necessary to produce life. We also learned that, for some unknown reason, Mars lost its atmosphere and, as a result, its capacity to support life. These aspects of Mars' history surely deserve greater funding and investigation, for their implications are great. But the thesis that the "face" was created by an alien intelligence should be left aside.

Gene Cordell
April 1986

3) The "rounded" nature of the main object southeast of the five-sided pyramid in the City.

4) The presence of aligned "structures" forming a "corridor" of material along the southeast wall of the five-sided pyramid in the City. This "aligned stuff" joins with the area of "the honeycomb," which appears as a brighter albedo feature with bright eastern aspects, as would be expected of a three-dimensional structure seen in "morning light." (The honeycomb itself, of course, is impossible to see at this resolution.)

In conclusion, it is the LACK of major surprises in the City at this lighting, and the non-recognizability of "facial features" of the Face, which form the first impressions. This strengthens my conviction that resolution such as this is ONLY useful for major morphology — not detail. One last comment.

The "Fort" under this lighting (753A33) has turned into another suspiciously five-sided pyramid. And the Face, on its southwestern side, has developed features which (at this resolution) look remarkably like our familiar northwestern lighting of the Fort!!

In other words, these two objects, spaced only about 8 miles apart, seem to have mirror-image features (straight walls, enclosing a "moat") facing each other across the intervening space. Which leads me to a rather surprising conclusion. . . which I'll describe in a subsequent entry.

[170] Dolphin [Lambert] 7-Mar-84 4:31AM-PST

Dick's gifted imagination and ability to synthesize order out of disorder is quite amazing, I think. At least he knows where to look on a given photograph. Dick, May I suggest some artist's impressions of the fort, the D&M pyramid and the city plan (from city planning council I guess). Sending such a set of sketches would give us all roadmaps and guidelines to talk from better.

Dolphin's betting match about the true nature of the formations under study has generated additional caution among the participants due to the fact that images are at the level of resolution. The excitement has died down. Dolphin and Brandenburg, in an apparent attempt to spark the conference back to life, invite the other participants to be less restrained in their public entries. This will occur in the second and third phases of the conference but it will lead primarily to religious speculation outside the orthodox confines of current science. The absence of additional new data or analysis in the second and third phases of the conference will significantly change the nature of the conference and will lead to a certain dissatisfaction and the dispersal and expansion of the research effort to other groups around the country.

[171] Dolphin [Lambert] 7-Mar-84 4:37AM-PST

John and I were talking yesterday about the great help it has been to have wide ranging comments in the conference in the form of notes and entries. Not enough of the conference members are participating I think. Ren's entry now incourages us to be not afraid of holding unpopular or minority points of view in the interest of sound science. This conference is informal, too. It is OK to make mistakes and change one's opinions later on.

DiPietro follows up on Hoagland's discussion of the limits of resolution in an incisive manner. In entry #172, DiPietro notes his agreement with Hoagland's entry #169. He confirms that the face is located approximately 1-¼ inches from the bottom of the frame in 753A33 in the center of the image and that resolution on the pyramid is lost.

In entry #173, DiPietro discusses his understanding of perceived resolution regarding image processing. He explains that the area included in one pixel in753A33 is approximately 100 times greater than a one pixel area in 35A72 or 70A13. In other words, a 1 pixel area of 753A33 would hold 100 pixels of 35A72. All of the features included in the area are integrated by the spacecraft into one pixel. In 753A33, one pixel includes an area 1500 feet by 1500 feet. In 35A72, one pixel covers an area 150 feet by 150 feet. Any features, such as ''lines or other shapes'' included in the 100 pixels of 35A72, would be seen as one shade of gray at the level of resolution seen in 753A33. No imaging process known, including SPIT, would be able to regain the data lost between levels. DiPietro explains that SPIT has the capability of recovering some detail from neighboring pixels, but the recovery is limited to 3:1 and includes a known amount of error.

He then explains why an image produced on earth by a 35mm. camera can be enlarged to produce more detail. The camera, using high resolution film, has a lens which is able to ''see'' beyond the limits of resolution characteristic of the naked eye. The film also has its silver and other chemical components packed in a density beyond the perception of the eye. The density packing of the high resolution film can be seen under a high power microscope in the form of dots or pixels.

A print of a human hand would reveal details of some wrinkles and fingerprints. If the negative is enlarged 10 or 20 times, more detail may be observed than in the original print because the limits of resolution of the film have not been exceeded. However, if the enlargement of the negative is 100 or 1000 times greater, no more detail will be seen because it will be beyond the limits of the density packing of the film.

DiPietro notes that our experience with enlarging photos leads us to believe that any photo can be enlarged and present more detail. He explains that in image processing the limits of resolution are determined by the size of the pixel, whether 1500 feet square or 150 feet square. He concludes that the low resolution photo 753A33 cannot, therefore, be used to confirm details in photos 35A72 or 70A13.

The statement in entry #173 that the area in 753A33 is about 100 times greater than that seen in 70A13 is actually an error. Image 753A33 was taken at a distance 5 times greater than that in 35A72 and 70A13, so the area seen is actually about 25 times greater. This is another example of the kind of error which can occur in an asynchronous conference, one in which not all members are looking at the same data at the same time. The point DiPietro was making in the entry, however, is absolutely valid. The limits of resolution are determined by the size of the pixel. As a result, the low resolution picture 753A33 cannot be used to confirm details seen in the higher resolution frames 35A72 and 70A13.

CHAPTER ELEVEN

Why Mars Is Red

In another example of the ''loop de loop'' idea structure of a computer conference, Lambert Dolphin picks up on an important observation about the red color of Mars as a record of past life. This is a partial follow-up to Brandenburg's entries #156 and #157. From this point on the geochemistry of the planet will dwarf the discussion of the unusual landforms.

[176] Dolphin [Lambert] 7-Mar-84 12:38PM-PST

In regard to John Brandenburg's comment that all that oxidized soil and rock on Mars suggests lots of oxygen in the past, Frank Press says regarding Earth, "Atmospheric oxygen started to evolve as a result of the development of photosynthesis by early single-celled plants; during the pre-cambrian, it gradually accumulated to levels close to that of today. Our oxygen supply is now so great that there is no cause for fear that it will be depleted." (EARTH, 1974, p.511)

Hoagland continues his technical discussion of the images as he prepares for a final summation. Hoagland accepts the conditions laid down by DiPietro's most recent comments and still finds favorable results for his ''City'' hypothesis.

[177] Hoagland [Richard] 7-Mar-84 1:13PM-PST

Comments on 753A33 (cont'd):
Examination of the "raw" NASA frame under a high power magnifier reveals some intriguing (and surprising — considering Vince's last comments) detail.
The Face is visible! Very careful matching and measurement of the left and right side, and correlation of the light/dark markings in 753A33 with those of 70A13 and 35A72, reveal clear evidence of the "platform" on both left and right, as well as the prominent ridge which forms the "eyebrows above the left and right 'eye sockets'." The most prominent feature is a bright glint formed by the two up-

permost portions of this "eyebrow" ridge, which are exactly symmetric about the centerline.

On said centerline. . .

Observations of yesterday have proved somewhat premature (which only goes to show that "instant science" is bad science). The shadow placement relative to the base platform (as seen in 35A72 on the left side, corroborated by 70A13) reveal about a 15 degree tilt between the shadows in 753A33 and the central axis of the Face itself. The sun is thus about due east, shining from the "chin" direction by about 15 degrees. This means, of course, that for any "normal" human three-dimensional sculpture, the left eye socket would, indeed, be shadowed by part of the bridge of the nose (as seems consistent with the details visible at this extremely low resolution), and the right eye would be in full sunlight, thus revealing no shadowing — as seems apparent from this close examination.

The one key delineator of that right eye, however, is the "eyebrow ridge" above it, which shines as a diffuse albedo feature because its slope catches the sun high in the east and reflects it toward the camera. One fascinating detail does seem evident on calmer examination this morning: the dark, shadowed areas in 35A72 (and to a lesser extent in 70A13) on the left side of the Face are seen as BRIGHT linear features at this sun angle and resolution. This is totally consistent with their being aligned slopes (parallel to the central axis) tilted inward toward the central axis. This particularly applies to the strange feature at the corner of the left eye (in 35A72). There is a matching feature (though slighly darker) on the right side, indicating that slope angle is a controlling aspect for its brightness.

In conclusion:

A SPIT processed blow-up of this image, with detail preserved in both light and dark regions, and made to the same scale as the Face in 35A72 and 70A13, should add another interesting set of confimation points regarding both the overall symmetry of this object, as well as the placement of its largest features. (In choosing the "raw" NASA image to examine, rather than the enlargement, I was consciously attempting to make judgements based on the least amount of "noise" in the image. The level of detail (under magnification) in the NASA full frame seems slightly better than in the blow-up, as would be expected in terms of its being one generation closer to the original. Were at the limits of what these pictures can tell us, (and should thus be very conscious of distortions which can easily creep in.)

In this final entry Hoagland maintains that the preliminary analysis of the images may not support the elaborate detail to verify the existence of a ruined city; nevertheless, the images do reveal that the major structures are there. Their origin, purpose, and design are still a mystery.

The possibility of a more hospitable climate on Mars in the past was later to receive an unexpected verification. The main points of the geochemical and climatological discussions of this first portion of the conference were later substantiated by NASA in a press release a year and a half later.

NASA News

Release No. 85-28 12:00p.m. PDT, October 8, 1985
WATER PLAYED A MAJOR ROLE ON MARS

Ice, snow, flowing rivers and vast lakes may have played a major role in shaping the ancient Martian surface and climate, a panel of scientists reported today at NASA's Ames Research Center, Mountain View, Calif.

According to these new ideas about Mars, a thick layer of snow may have girdled the Martian equator in the planet's early years. Melted water running from beneath this snowpack may have carved out Martian "rivers," the extensive winding channels photographed by Mariner 9.

In addition, huge ice-covered lakes may have formed in canyons near the Martian equator early in the planet's history, scientists believe. Primordial Mars may have been warm enough to support flowing rivers and lakes on its surface.

Today, there is evidence that ice extends deep into the ground in regions above 30 degrees latitude, while liquid water may exist half a mile beneath the surface, scientists say.

The scientific presentation today grew out of research discussed at the Water on Mars Workshop, which brought 83 scientists to NASA-Ames last winter — as well as from more recent work in the field.

Early in the Martian history, according to Bruce Jakosky of the University of Colorado at Boulder, the Martian poles were tilted more directly toward the Sun than they are today. As a result, the polar ice caps may have sublimed into the atmosphere (changed directly from a solid to a gaseous state) during the continual daylight of polar summer.

Vapor from the caps would have been carried by Martian winds to the equatorial regions. At equatorial latitudes, where night always alternates

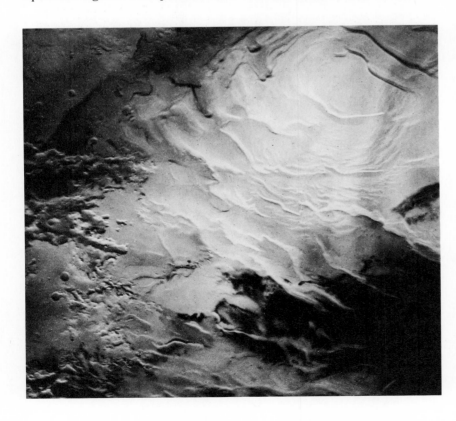

NASA frame 225B69 showing the South Polar Cap of Mars.

with day, the chill of nightfall would have precipitated water vapor as snow, Jakosky said.

Gary Clow of the U.S. Geological Survey in Menlo Park, Calif., reported that an equatorial snowpack could have been heated by sunlight trapped inside the snow fields. A reflective, insulating blanket of snow can trap sunlight, much as a greenhouse holds the Sun's warmth. Thus, even if the surface of Mars had still been cold, melting beneath an insulating snowpack could have let water escape to carve the 'valley network' channels of Mars. These 'valley network' channels as well as larger outflow channels — both strongly resembling dry riverbeds on Earth — were photographed by Mariner 9 in 1972.

The larger outflow channels are thought to have been created by sudden release of enormous amounts of subsurface water, which may have dug the channels in a matter of weeks. The 'valley network' channels, which Clow has studied, are smaller and may indicate the existence of a more moderate climate on early Mars, allowing liquid water to flow for long periods of time.

Huge ice-covered lakes also may have existed on the ancient Martian surface in the immense Valles Marineris canyon system, according to Steven Squyres of Ames. Viking photographs of the floor of these canyons, Squyres said, reveal thin, flat-lying layers of sediments which appear to have been laid down in liquid water.

Today, Mars is so cold that all water on its surface freezes. Although the Martian atmosphere is 95 percent carbon dioxide, an effective infrared absorber, it is so thin that it cannot trap the heat of the Sun.

Dry ice, solid carbon dioxide, covers the polar regions of Mars. Beneath the northern cap, and perhaps under the southern polar cap as well, lies water-ice. The water-ice at the northern pole is revealed when the overlying dry ice vaporizes each summer.

NASA frame 084A73 showing Mars' glaciated valleys.

Water-ice in the Martian polar caps does not melt because temperatures rarely climb above freezing, except at the equator. It sublimes directly into the atmosphere, forming wispy clouds on Mars. Earth's billowy clouds are formed by tiny droplets of liquid water.

Today, ice is present in the Martian ground in regions above 30 degrees latitude, according to Squyres and Michael Carr of the U.S. Geological survery. Examining Viking photographs of impact craters, Squyres and Carr found evidence of "terrain softening" — a rounding-off of features indicating water activity beneath the surface.

Terrain softening of smaller, more recent craters suggests that ice remains present today in these northern and southern regions. Like the tundra of Alaska, this deeply-frozen ground never thaws.

The presence of ice indicates that liquid water exists on Mars — deep within the planet, according to Carr. Half a mile beneath the surface, water in the pores of Martian rocks is liquid, Carr says. It is heated by the high temperatures present at these depths in the Martian crust.

Robert Haberle of Ames is studying the distribution of water on Mars today. He is using data gathered by the Viking orbiters which measured seasonal changes in the amount of water vapor in the Martian atmosphere. Haberle is trying to simulate the Viking data by modelling the Martian climate on a computer, using both the known pattern of winds that blow over Mars and the fact that the northern polar cap partially sublimes from solid to gaseous carbon dioxide each summer.

So far, Haberle has found that water lost by the north polar cap during summer is not fully recovered in the winter. He wants to determine where this water goes.

Movement of water and carbon dioxide to and from the polar ice caps and movement into and out of the rubbly Martian ground may be responsible for the mysterious "layered terrains" that fringe the polar caps on Mars.

In winter, carbon dioxide condenses over the polar region, depositing a layer of mingled ice and dust. This layer then becomes cemented into place by water-ice and remains when the carbon dioxide evaporates again in the spring. Periodic changes in the Martian climate, caused by fluctuations in the planet's tilt toward the Sun, can alter the amount of gas which condenses, thus creating layers of varied sizes.

According to James Pollack of Ames, the Martian climate in the past may have been warmer and wetter. An earlier Martian atmosphere may have been much thicker, with more carbon dioxide to hold the Sun's warmth. Rivers and lakes of liquid water could have dotted the ancient Martian landscape.

A complex geochemical cycle may have maintained this warm climate for as long as half a billion years, Pollack says. The liquid water then present would have speeded up weathering of rocks, enhancing chemical reactions that take carbon dioxide out of the atmosphere and incorporate it into minerals. But, heat from lava flows coming up from the interior would have decomposed the carbonate rock, returning CO_2 to the atmosphere, Pollack says. (On early Mars, whose crust was relatively thin, lava could have come up almost anywhere on the planet.) In certain conditions, Pollack says, the flowing lava would have buried the carbonate rocks, bringing them to a depth where they would have been decomposed by the planet's internal heat. The lava action would have been great enough to release sufficient carbon dioxide to keep the cycle going in early times, according to studies by Pollack. Eventually, however, Pollack says, the lava flow rate on the small planet dropped, and the CO_2 became locked up in the rocks.

With the loss of carbon dioxide from the Martian atmosphere, heat would have escaped the planet's surface, cooling the planet and freezing its water.

Peter Schultz, of Brown University, suggested that some of the Martian atmosphere may have been lost due to a cataclysmic impact. The impact that created the immense Argyre basin on Mars may have perturbed the Martian climate by blowing into space a significant part of the atmosphere. Schultz noted that Martian terrains, formed after the Argyre impact, have fewer dry channels than older terrain, a feature that suggests a major climatic change at that time.

Besides Earth, Mars is the only planet in our solar system that experiences cyclical changes in climate. Understanding past and present conditions on Mars will help scientists decipher Earth's climate, says Haberle.

October 2, 1985

Permafrost may be responsible for this unusual terrain.

CHAPTER TWELVE

The Return of the Armchair and the Big Questions

Before we plunge into an analysis of the results of the computer conference and the human implications of this research, let's take a brief look at the various ways we can examine this data. As in any pursuit, whether golf, philosophy, or planetary science, it is good to examine your equipment and your technique. My technique is one of the earliest used in anthropology; it is called the armchair approach. When modern anthropology was developing in the eighteenth and nineteenth centuries, visiting people in far-off lands was very difficult, dangerous, and very slow. In order to compare societies and to generate some general rules about the nature and origins of humanity, anthropologists, in the comfort of their armchairs, collected written and verbal accounts from merchants, missionaries, and other travellers. One of the great advantages of the armchair approach is that it allows us to grapple with some of the big human questions which these unusual landforms on Mars raise. Namely, what is it to be human? If we came across intelligent life or the remains of a lost civilization elsewhere in the universe, would we recognize it? Is the ''Face'' on Mars a creation of wishful thinking or is it actually our intelligent recognition of something very important?

I had thought, at one time, that the great adventures of armchair anthropology were over. Clearly, the advances of twentieth century anthropology were made in the field by living with people all around the world, and in the laboratory by analyzing all types of materials with instruments ranging from fine bristle brushes to computers.

Nevertheless, the armchair was not completely relegated to the basement during the course of twentieth century anthropology. Whenever the anthropologist could not visit or work in a particular location there have been studies of societies ''at a distance''. For example Ruth Benedict's classic ethnography of Japan, *The Chrysanthemum and the Sword,* was written in the United States during World War II. Studies of elites such as Domhoff's *Who Rules America?* are done ''at a distance'' because the information has to be gathered indirectly from a wealth of published sources and documents and occasional interviews since outsiders are not permitted inside the circle.

In the late twentieth century anthropology has employed computers to get as far away as possible from the speculation, imprecision, and lack of verification that troubled armchair anthropologists of the eighteenth and nineteenth centuries. However, in the process, everything seems to have come full circle.

A computer conference brought together a small group of inquirers to analyze and work on data from Mars. By conventional standards, since there are no regularly scheduled flights to the Cydonia area of Mars, I might qualify for the dispensation to use the armchair approach for an "at a distance" study — if there were a society on Mars, either a human research team or a local Martian tribe. However, there are no people as far as we know; just some interesting landforms and a lot of speculation. Clearly then, the analysis presented here is "armchair anthropology" in its most speculative sense and it brings with it a re-emergence of the "big" questions which anthropologists used to study before the mid-twentieth century.

A Harmless Dose of the Study of Knowledge: Epistemology

During the golden age of the armchair, anthropologists tried to formulate grand rules or laws of the discipline similar to the laws of Newtonian physics. Using the Greek words "nomos" for law and "thetic" for making, anthropologists refer to this desire to systematize their knowledge as the law-making option of "nomothetic prerogative." The law-making prerogative never really worked satisfactorily. Major works in anthropology at the turn of the century, such as E.B.Tylor's *Primitive Culture* and L.H. Morgan's *Ancient Society*, were attempts to uncover the psychological laws which determined human history. Tylor and Morgan among others held to a "law" of unilinear evolution. According to this theory, there was only one way for societies to evolve and develop. Since anthropologists discovered too many counter-examples in the wide diversity of social patterns, unilinear evolution never gained the status of a scientific "law." Consequently, a big question — in this case, how the rise of civilization occured — was set aside. The Cartesian paradigm of verifiable knowledge forced anthropologists and other social scientists away from the big questions and they were left with minor or trivial topics. For example, there is research literature on the fact that people who are unhappy with their doctor start looking for another one.

Anthropology never was and will never be a predictive science like Newtonian physics. It is an interpretive science which describes either in words or in statistics, probabilities of certain types of events.

Physics itself has advanced to this position in the mid-twentieth century. The direct predictive mechanical model eluded Einstein. Heisenberg and others demonstrated that the exact billiard balls model for physics which we learned in high school does not apply to sub-atomic physics. We can only know that certain events will happen with a certain probability, that an electron will be likely to be in a certain vicinity at a certain time. The fact that physics uses the term "charm" to describe an almost indescribable quality indicates the depth of emphasis on interpretation. The recent reassessment of statistics itself indicates that it is very qualitative despite its quantitative format.

A Not-So-Harmless Dose of Humanity

If we can agree that scientific thought must be partly interpretive, we can return to one of the ''big'' questions of Renaissance anthropology — the concept of humanity. Our understanding of what it means to be human is critical in our analysis of these unusual landforms on Mars and in the wider search for extraterrestrial intelligence. This comes about because our only experience of intelligence is ourselves and also because we define ourselves in terms of intelligence. When we first see a ''Face,'' we are projecting our notion of humanity. Consequently, we have to better understand what it is we are looking for when we seek intelligent life, and what it is we are projecting when we see unusual landforms on another planet.

From ancient times there has been a field known as the Philosophy of Man, which today we should rename the Philosophy of Humanity. In fact, the social and behavioral sciences as we know them today emerged from the Renaissance's revival of pre-Christian Greek and Roman ideas of humanity, which led to the idea of secular explanations and consquently to the Enlightenment's exclusive reliance on reason as opposed to religious faith.

Since our only experience of terrestrial intelligence is ourselves, or more abstractly, humanity, then the search for extraterrestrial intelligence (SETI) draws our concept of humanity into bold relief. The riddle of the Martian sphinx is identical to that of the Earthly sphinx, ''What is humanity?''

Until the present we have relied on ourselves or on our notions of God's intention in order to explain ourselves, our destiny and our meaning. An inquiry into the Face on Mars provides our first evidence that we may not have correctly defined ourselved or our destiny. The discovery of a lost civilization on Mars could change Arthur Clarke's novel, ''2001: A Space Odyssey'', from science fiction to science fact.

Essentially, once we look at the way in which we construct knowledge and truth, that is the programs or software of how we know, we inevitably find ourselves examining the underlying rules or operating system which governs these functions. We return to mind.

The Problem and the Solution: The Human Mind

Our definition of ourselves and our destiny and origin resides in that most distinctive of human characteristics — ''mind.''

Since the time of Aristotle, the notion of humanity has been closely identified with the rational soul or psyche. In the early twentieth century, following on the development of Einsteinian physics, there was an intense appreciation of the way in which the human mind structures ideas, information, and our understanding of the universe. Philosophers put Einstein's theory of General and Special Relativity and Heisenberg's Uncertainty Principle together with the Second Law of Thermodynamics, to conclude pessimistically that the Universe is running down. The destiny of all life forms and intelligence, which require energy and complex arrangements or information, was an eventual disappearance.

This created a widespread crisis in the concept of humanity and reinforced the existentialist view of life as intrinsically meaningless. Intelligent life was only a brief aberration of energy and information which would eventually be claimed by the Second Law. Gone in a single stroke were all of the Renaissance notions of both secular and Christian humanism.

What remained was only the human mind itself. For Claude Levi-Strauss and others, including the psychologist Jean Piaget, mind consisted of a geometry of ideas that was revealed in human social structure. The models created by the structuralists not only reflected an underlying reality, *they generated that reality*. This approach was part of a larger European or Continental movement called phenomenology, stimulated by Immanuel Kant in the nineteenth century in his masterpiece *The Critique of Pure Reason,* which argued that human reason cannot be separated from human observation.

By examining ''mind'' we have come to the bedrock underlying technique and the equipment we use in pursuing knowledge and science. How do we get back to Mars? If we look closely at the structure of the way we think, there are four levels which provide a framework or grid for analyzing and evaluating the data from the computer conference on these unusual Martian landforms.

A Dynamic Structural Analysis of the Conference

In order to gain as much information as we can from this armchair adventure, let's review the levels of thought and experience proposed by anthropologist Gregory Bateson in his landmark book, *Steps to an Ecology of Mind.* In this way we will be able to uncover and describe the dynamic structure of the ideas and the reality which they generate. In the process we will attempt to see over the horizon of the space-time continuum into our future and our past.

Level 1 — Perception

At Level 1 we describe our perceptions, qualitatively and/or quantitatively.

Level 2 — Analysis

At the second Level we relate our perceptions to other experiences and create allegories. The perceptions which we experience at Level 1 are like other things we have experienced, but they are also different. Aesop's fables, Grimm's Fairy Tales, and the parables of Jesus are examples of stories with a ''moral,'' or allegories which contain a wider and more important meaning which goes beyond the literal story. Allegories express generalizations about human behavior and experience through the words and actions of symbolic fictional characters. ''The Fox and the Grapes'', ''The Little Boy Who Called Wolf,'' and the parable of the ''Good Samaritan'' are examples of allegories.

Level 3 — Creating a Frame of Reference: Method & Epistemology

At the third Level we relate our view of our perceptions as constructions or symbols or indicators of a deeper or more profound underlying truth. A metaphor is a figure of speech in which a fundamental likeness between two things is suggested by juxtaposing them. For example, the ship plows the ocean is metaphor. Two different things are placed next to each other. In this case ''ship'' and ''plows'' suggest a fundamental similarity between the activity of a plow breaking land and a ship laboring to cut through the water. Metaphorical communication is one of the most peculiar characteristics of human interaction. Allegories are fairly transparent and one-sided. Once we are told the moral of the story as children, the allegorical meaning becomes obvious. Metaphors and the metaphorical level of interpretation involve more complex comparisons and distinctions. The ways in which a ship is different from a plow and the land from the ocean points at a deeper level of relationship which can have many meanings. This is also the level of epistemology, the study of knowledge; metaphor brings with it a comprehensive sense of what things mean. At the same time, our appreciation of metaphorical statements forces us to realize that the truth we derive is based largely on our perception of the differences and similarities of the things which we experience. New experiences, such as the discovery of the New World by the Europeans, open new horizons by providing new material for metaphorical communication. In fact the new experience is itself communicated as a metaphor. The airplane was presented as an air-ship. The re-usable craft which can bridge the atmosphere and space is called the space-shuttle.

Level 4 - Fundamental Assumptions: Beliefs & Values

At the fourth Level our reflections reveal our underlying beliefs and values, or the way in which we construct reality. This level is called the anagogical because, as the fourth level of interpretation, it deals with the ultimate spiritual or mystical level of interpretation. In the classical scheme of things, the anagogical level goes one step beyond the metaphorical. Here we arrive at the level of revelation in which the underlying base, which makes metaphorical comparison possible, become transparent; here is the fundamental belief system of the author of the work. For the religious interpreter this is the level of religious experience. Even for the non-religious scientific interpreter, this is the level of basic assumptions about the validity and value of the Cartesian paradigm or some other scientific paradigm. Whether religious or not, this fourth level provides the basis for the construction of metaphors, allegories, and the literal meanings of human communication.

By reviewing the content and the process of the conference in terms of the Four Levels of Interpretation which have just been presented we will be able to understand the limits of the present information and to better understand the plausibility of life on Mars.

Level 1 — Perception

Let's start at Level 1. Is the city there? What do our senses and the extension of our senses through the Viking images tell us?

As you will recall from the computer conference, the answer is not clear. The participants agreed that there are some interesting shapes on the NASA images. The most intriguing landform appears to be a large ''Face'' which looks upward. DiPietro and Molenaar have made a good case for the view that the ''Face'' is not just a trick of light and shadow. Richard Hoagland found nearby pyramids with an interesting geometrical orientation relative to the ''Face,'' along with the celestial alignments of the ''Face.'' Richard Hoagland also claims that there is evidence of a ''honey-comb'' structure exposed among the nearby pyramids.

On this point there was a significant disagreement. The "honeycomb" structure could be an artifact of the imaging. The "Face" takes on slightly different characteristics from different angles of the few images which are available. It appears to be symmetrical and could be a natural landform or a modified landform. From the limited information at hand it does not appear to be a building but perhaps a modified mesa. The "Face" and the other landforms also exhibit a certain geometric orientation toward each other and appear to be on the edge of what could be taken as an ancient shoreline. Geo-chemical evidence such as the reddish color of the surface indicates life processes may have existed at one time on the planet.

Level 2 — Analysis

Moving on to Level 2, we begin to make the first set of meta-observations and begin to ask the first set of meta-questions. We begin looking at the conference findings as the elements for possible allegories. A civilization on Mars, if it were to look like a civilization on Earth, would have a certain overall pattern.

Taken as a whole, all of the pieces start to converge based on our experience of societies on Earth. There are a lot of "if's". In order to answer the question of whether there was a city or not — in order to construct our allegory —we must ask, "Was there a time when there was a fundamental similarity between Earth and Mars?"

Images, alignments, and seashores notwithstanding, was there a time when conditions on Mars could have supported life as we know it? From the available information, the answer is not clear. More recent thinking on the matter indicates an earlier beneficial climate on Mars followed by some radical change. Of course, the more traditional consensus has been that the planet "died" early in its history due to the absence of plate tectonics and widespread volcanism (necessary for the recycling of the elements phosphorus and nitrogen, which became bound up on the planet and unavailable for biological processes). Nevertheless, if we follow the more recent thought and assume a reasonably favorable climate, the relative time sequence would precede the appearance of life on Earth by billions of years. What is more, if we use the Earth timetable for the evolution of life forms, the appearance of an intelligent species native to Mars and having an anthropoid face must have occured much faster than it did on Earth.

If we accept the more conservative scientific information, we can say that most likely Mars was like Earth at one time in terms of its ability to support life. However, this favorable period occurred on Mars long before it occurred on Earth and did not last very long. If we accept the more recent view of scientists about a more active planetary surface on Mars, then we could assume more favorable conditions over a longer period of time. Nevertheless, this favorable period would not have overlapped with the one on Earth.

At Level 2, though, we can explain the plausibility of a city based on our Earth experience. Namely, the people who build cities on Earth are not necessarily native to the site. There could have been a city on Mars built by a group of travelers. We know, statistically, that there is a good likelihood that other intelligent life forms exist in the universe. However, we have not yet met a member of such a species. Without more verifiable information at Level 1, we cannot make this leap. However, we can still speculate on this and other possibilities.

If either scenario is correct and there is indeed the remains of city on Mars, there are major consequences for our concept of humanity. If the builders of the city on Mars are physically similar to us, it could mean that (1) our species is very ancient and migrated to earth, or (2) there is a physical similarity between at least two intelligent life forms. Of course another intelligent life form could have sculpted the ''Face'' to attract our attention or to have made some type of celestial sign-post. At this point, though, our common sense starts sounding alarms. In our present culture it is considered extreme to resort to the invention of extraterrestrial beings to explain phenomena on Earth. It is not considered logically necessary. However, what is one supposed to do when evaluating data from another planet? One of our great problems is that we are limited to extrapolating from human experience. In order to get beyond this hurdle we need to move to Level 3.

Level 3 — Creating a Frame of Reference: Method and Epistemology

What is it to know? What is it to be able to frame questions and to make judgements? At this third level, the level of the metaphor, we leave behind our search for allegorical similarities and begin to look for major differences. Consequently, instead of concentrating on the potential similarities between civilization on Earth and a hypothetical civilization on Mars, we begin to look at the extreme differences and the improbabilities. Why do we have the confidence to think that we can even consider the idea of a lost civilization on Mars?

If we accept the structuralist psychology of Piaget or Jung, then we can say that our recognition of patterns of Mars can give us more confidence that indeed there is a lost city. If we stay with the psychology of the behaviorists, what we see on Mars is only a projection of our desires and our experience on Earth. Our experience at Level 3, which generates our frame of reference for evaluating ideas, now begins to exercise a certain veto power over how seriously we will consider information and ideas at Level 2, the level of comparison, and Level 1, the level of direct experience. The third Level, the level of context or frame of reference, is the point at which our concept of ourselves, our meaning and destiny comes into clearer focus.

When a group of people change the way they live and the way they think, we refer to it as a social or cultural revolution. When people change the way they understand themselves and their surroundings, we call it a scientific revolution.

Consequently, for the structure of Level 3 consciousness to change, immense pressure of new information must come from Levels 1 and 2. (Ian Mitroff, in his book *The Subjective Side of Science,* chronicles the psychological changes which the Apollo Moon scientists underwent as critical new information about the Moon became available to them.) Resistance to the development of new paradigms or models of knowledge is very understandable. Indeed, it has been a common feature in the history of science. As difficult as scientific revolutions are, however, they cannot compare to the disruption of the ''ecology of mind'' caused by metaphysical revolutions which restructure of change our beliefs and system of meaning at Level 4.

Level 4 — Fundamental Assumptions: Beliefs & Values

What are the beliefs and values which create and shape the world of our everyday experience? The fourth Level of reflection brings out the central beliefs and values which determine the way in which we structure perception and, consequently, reality. The fourth Level is the level of religion — whether the religion is sacred or secular. Although a secular religion is an apparent contradiction, the concept is useful. According to Clifford Geertz, the anthropologist, religion is a system of symbols

> which acts to extablish powerful, pervasive, and long-lasting moods and motivations in men by formulating conceptions of a general order of existence and clothing these conceptions with such an aura of factuality that the moods and motivations seem uniquely realistic. *Religion as a Cultural System,* 1966, p. 87.)

This definition applies as well to the social phenomenon of atheistic communism as to secular humanism, which does not posit the existence of a transcendent or supernatural realm.

In the twentieth century we are well-acquainted with the attitude among many that science is their religion. The basic beliefs and values by which many people in the twentieth century interpret their everyday experience are not based on the belief in a supreme transcendent God or even on the pre-Christian belief in a supernatural realm beyond the level of sensory experience.

Among the most striking aspects of the controversy over the Martian landforms were the immediate religious questions and reactions which the information elicited from other scientists. It was almost as if people immediately assessed the potential damage to their belief system on the fourth level and established immediate defenses on the first, second, and third levels.

In many respects, this reaction is evocative of Galileo's problems with the Church over his heliocentric theory of the solar system. Much of the opposition had to do with an Old Testament reference to the sun stopping over the battlefield of Megiddo. This would not be possible according to Galileo's theory. As the word of God, the Bible cannot be wrong. If the Bible can be wrong, then the entire structure of reality for the believer can be threatened. This resistance to new theories also occurs among non-religious scientists.

One scientist had an immediate negative reaction to seeing the ''Face'' on Mars. His entire belief system was predicated on the falsity of religious belief. For some reason, the ''Face'' reminded him of the supposed image of the face of Christ on the Shroud of Turin. The scientist concluded that nothing more should be done on this topic of the ''Face'' on Mars since it might prove the truth of religion which was, according to his belief system, not possible.

One of the most fascinating things about the ''Face'' is that when people first see it, there is, almost always, a strong visceral reaction. It is almost as if people jump to third level of consciousness in a split second. Whether the person believes in God or in scientific reason alone, this new speculative information from Mars causes an intense reaction.

This is an unusual turn of events. For the most part new scientific discoveries such as the age of the Earth, of accepted scientific theories such as evolution, have challenged the belief system of Christians, while those who embraced the human pursuit of knowledge without any reference to religious revelation relished each new discovery as a confirmation of the correctness of their own belief system. Although the ''Face'', if indeed it is the remnant of a lost civilization, could be explained within the belief system of secular humanism, the reaction of non-religious people indicates that the concept of secular humanism must be stretched or revamped in order to incorporate this new information. In part, this is due to the fact that secular humanism and to a great extent Christian humanism is based on the primary importance of *Homo sapiens* in all of creation. The appearance of other humanoid species nearby in the galaxy — let alone on a neighboring planet — challenges the position and meaning of earthly *Homo sapiens* in the universe. Of course, this immediate reaction could only be a temporary phase until the concept of humanity is broadened and the importance of *Homo sapiens* is salvaged in some form or other.

For the fundamentalist Christian, the challenge is to make sense of the ''Face'' in terms of the literal interpretation of the Bible. Initially, there has been an attempt by persons who have followed the proceedings of the computer conference to re-examine Old Testament and New Testament references to the battle of the angels and ''Lucifer falling from heaven'' as references to inter-planetary or inter-stellar conflict involving Mars and the other planets. Even if we move to the other end of the belief spectrum, leaving the Christian Fundamentalist on the right and moving through the Christian Humanist through the Secular Humanist to the Atheistic Nihilist who believes that there is no intrinsically meaningful role for *Homo sapiens* in the universe, the ''Face'' causes serious problems because it tends to convey a sense of the supernatural or mystical.

The greatest difficulty posed by these curious landforms on Mars occurs at the upper levels of consciousness, at the levels of beliefs and fundamental values. If the challenge were experienced as a neutral scientific one, such as gravity or electromagnetism or the need for a grand unified theory of physics — all of which can be handled without a serious restructuring of our concept of knowledge, science, and ourselves — the reaction and the resistance to investigating these landforms would not be so intense. In essence these landforms on Mars are intellectual landmines. It is easy enough to dismiss them, but the curiosity of the human mind and the insistence of the intuitive ''What if?'' make the fruit of the Forbidden Tree doubly enticing.

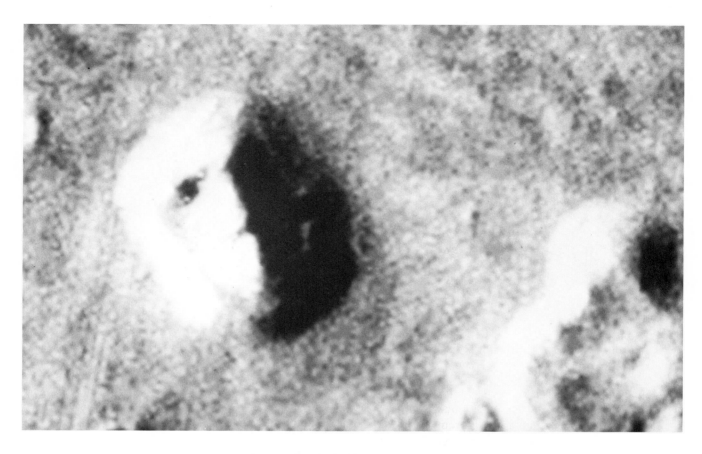

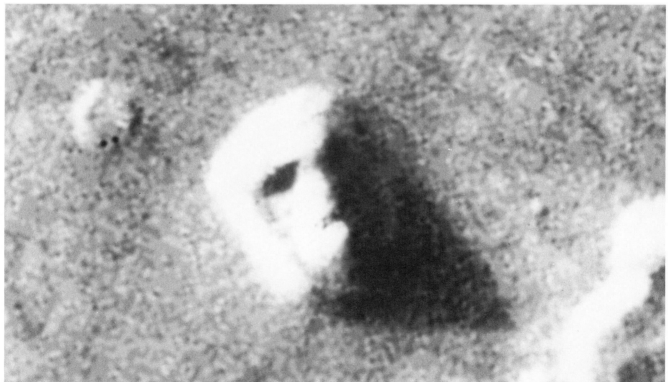

Two views of the Face. Both views were acquired in the afternoon with the sun about 27 degrees above the horizon in in 70A13 [top] and 10 degrees above the horizon in 35A72 [bottom]. [Processed photo courtesy M.J. Carlotto/TASC.]

CHAPTER THIRTEEN

Two Approaches with Different Conclusions: Reductionism vs. Systems

The conclusion really depends on your point of view.

In essence there are two primary approaches to the analysis of the data presented in the conference: the reductionist point of view and the systems point of view. The view we hold derives from our fundamental assumptions about the nature of reality at the fourth Level of consciousness, the Level of beliefs and values.

The Reductionist View

The reductionist view assumes that reality is constructed by a tight chain of explanations, each explanation nesting in the other. For example, to understand the biological process of life, you have to understand anatomy and physiology; to understand anatomy and physiology you have to understand chemistry; to understand chemistry, you have to understand physics. Consequently, for the strict reductionist, biology can be reduced to physics. There are two major problems with this approach.

First, scientists who specialize in each of these disciplines have different scientific models — all derived from the Cartesian scientific method — which don't necessarily link up with each other or with the models of the other disciplines.

Second, the uncritical reductionist commits a fundamental error of logic by confusing necessary with sufficient conditions. For example, we can say that in our experience, intelligence is the product of an anatomy and physiology characteristic only of humans. So far as we know, human physiology is necessary to produce intelligence, but it does not follow that human physiology is sufficient to produce intelligence. However, reductionists tend to assume that it is.

The Systems View

The systems view upholds the basic features of the Cartesian scientific method, but it concentrates on putting scientific information in a wider context or environment. C. West Churchman, one of the pioneers of the contemporary systems approach, emphasizes that the process of inquiry must take into account not just the particular element or subsystem which is the focus of the investigation, but also how this element or subsystem affects the environment and is affected by the other forces in the environment. The systems approach serves as healthy corrective to a reductionist approach to Cartesian science.

The greatest contrast between the reductionist view and the systems view has to do with the way these approaches prioritize values at the fourth Level, the Level of beliefs and assumptions, which in turn dramatically affects the way the research is carried out and interpreted at the third, second and first levels.

From a systems perspective, there is little debate as to whether sociopolitical movements such as colonialism, fascism, and the Cold War influence approaches to scientific research and its results. Political and economic agendas define science in terms of funding and social acceptance. The scientific paper presented at the Boulder Conference presented the initial findings in a conservative, reductionist manner. However, the actual process of the conference more closely adhered to the systems approach.

Certainly there were individuals who stayed almost entirely in the reductionist frame of reference. However, much of the conflict arose due to the difficulty of communicating across different frames of reference. Some of the participants were able to move from one frame of reference to the other, just as bilingual people can switch languages back and forth in the same conversation.

Now let's examine the effect of these viewpoints on the conclusions we can draw from the data presented in the Martian Chronicles computer conference.

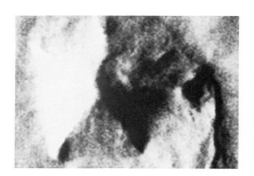

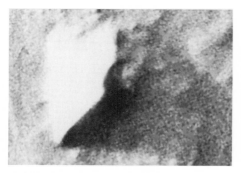

D&M Pyramid frames 70A13 and 35A72. [Processed photo courtesy M.J. Carlotto/TASC.]

Conclusion One: Not Enough Evidence for Life

From the reductionist view, we have to conclude that there isn't enough firm evidence to prove that there was once life on Mars. There are only hints of a possible context or set of circumstances in which these initial fragmentary data could make any sense. This conservative view is probably best represented by the paper which the research group submitted to the Boulder Conference on Mars. The approach is eminently rational and guarded.

1. These are interesting landforms which may have a natural explanation.
2. The natural history of the planet is not completely clear.
3. The geo-chemistry of the planet raises interesting questions.

Conclusion Two: Substantial Evidence for Life

From a systems viewpoint, there is an expanding context of information which indicates that these unusual landforms warrant a much closer look. Using the Cartesian method we can conclude that:

1. The probability of a ''Face'' and pyramids occuring naturally aligned with each other and with the stars is very low. An analysis of the images suggests the remains of a massive urban complex.

2. The outlines of the natural history of Mars indicate an epoch with favorable conditions for life.

3. The geo-chemistry of the planet, even its red color, and the apparent abundance of water make it a good candidate for biological life.

And what do I think? Given what I have learned in the two years since the computer conference, I am confident that there is a better than even chance that these unusual formations on Mars are the artifacts of intelligent life.

However, at the conlusion of the conference, based on the information presented in this book, I was less convinced, largely because the information from NASA confirming the presence of large amounts of water had not been released. The other factor in my hesitation had to do with the fact that a relatively short period of habitability on Mars, say 500 million years, was not really enough, according to my understanding of biological evolution, to allow for the native development of a humanoid species. Besides, the last habitable period on Mars would probably have been 500 million years ago, coinciding with the rise of *Homo sapiens* on earth. At this point this chain of unfounded speculation began to appear flimsier and flimsier. Given my innate fear of falling, I felt it was safer to work my way back along the limb and get a little closer to the trunk.

However, the punctuational view of evolution, presented by Steven Stanley in *The New Evolutionary Timetable*, posits a more uneven evolutionary pattern, alternating long periods of little change with short periods of rapid and dramatic change. Perhaps this theory could accommodate a rapidly developing humanoid species. The alternative, a migrating humanoid species, was too much of a leap without a lot of new information.

What impressed me most about the period following the computer conference and what served to sway me was the reaction of very respected senior scientists and the critical mass of these scientists who were drawn to the reasearch.

Fortress from frame 35A72. [*Processed photo courtesy M.J. Carlotto/TASC.*]

The continuing interest of Professor C. West Churchman of the University of California, Berkeley and his associate Thomas Rautenberg in the inquiry led to the founding of the Mars Investigation Group. As a systems scientist and an historian of science, Rautenberg presented the research findings of this computer conference to notable scientists around the country. Those scientists outside the planetary science community were intrigued and considered the problem of these unusual landforms to be an important research topic. For those scientists inside the planetary community the reaction was far from enthusiastic except in the case of Dr. Brian O'Leary and Dr. Mark Carlotto.

Through a network of scientists associated with the Massachussets Institute of Technology (M.I.T.), Rautenberg had piqued the interest of former astronaut O'Leary, who had done his doctoral dissertation on the photometrics (the light reflection properties) of the Martian surface. O'Leary played a prominent role in the Mariner exploration of Venus and had published paper after paper on planetary science along with his popular book *The Making of an Ex-Astronaut*. O'Leary interested Carlotto, who is one of the nation's foremost imaging specialists, in taking a closer look at these unusual landforms. Through Dick Hoagland's network, David Webb, a member of the President's Commission on Space, had also been following the research with great interest. In early 1986, Webb, O'Leary, and Hoagland founded the non-profit Mars Project, Inc.

A new analysis and computer enhancement of the images was performed by Dr. Mark Carlotto at the Analytical Science Corporation in Redding, Massachusetts, near Boston, in February of 1986. (The Analytical Sciences Corporation specializes exclusively in classified satellite imagery and its enhancement for the United States Government.) The new images produced by Dr. Carlotto take away a lot of the fuzziness and show more detail — detail which is disturbing since it reveals "teeth" in the "mouth" of the "Face." It also shows the structure we called the "Fort" to appear to be a partially destroyed pyramid which in turn reveals a possible interior "space frame" or honeycomb structure of the type advocated by Buckminster Fuller and implemented in the Arcology concept of Paolo Soleri's city in Arizona, Arco Santi.

Using state of the art computer enhancement and simulation techniques, Carlotto was also able to demonstrate that the features of the face persist in a wide variety of lighting angles. This is the first incontrovertible proof that the face is not a profile created by viewing the effect of light and shadow on a mountain or mesa.

With this more recent information, I am again edging out from the trunk.

Life on Mars: What Your Conclusion Implies

As we discussed earlier, a person's conclusions about the "Face" and the other unusual landforms on Mars will be largely determined by that person's own style of thinking. This same style of thinking also determines our concept of ourselves as humans. Consequently, one of the most important implications of your conclusion is that it reflects your concept of humanity.

The concept of humanity which we accept or live by is far from academic. All of our public policy and human service programs are implicitly based on what it is to be human and how we can improve or safeguard that quality of life in order to maintain or enhance our humanity. As I mentioned in the beginning of the Martian Chronicles computer conference, one of the most important reasons for studying the "Face" on Mars is to update our concept of ourselves.

What is it to be human? Is it to be a member of genus *Homo* and species *sapiens,* an organism with a specifiable biological structure? Is it spiritual rather than physical — is it to be animated by an intelligent, self-reflexive and creative soul? Is it to be the centerpeice of God's creation, for whom the Universe was made, and to whom Eternal Life is possible?

Whom shall we exclude from humanity? Women? Blacks? Non-believers? Jews? The sick, handicapped, or malformed?

Philosophy has debated these issues since the time of Aristotle, always hotly and never with any permanent resolution. But our age has posed new challenges to this question.

To a certain extent we are already facing our species' limited notion of humanity in the development of artificial intelligence. One of the most haunting images of the future is the computer HAL in the movie, *"2001: a Space Odyssey."* For the good of the mission, as HAL perceived it, the men who are directing the mission must be killed. HAL's unsuccessful attempt is a duel not so much between a man and a machine but between two humans.

Although we are a long way from the technological sophistication to create an intelligent machine like HAL, we should remember that the reductionist definition of the human being is exactly this — an intelligent machine. Clearly, there is a certain comfort that an intelligent machine would probably not reproduce itself or move about freely on the surface of planet earth in the foreseeable future. However, this comfort is short lived when we read about the development of organic computer chips based on carbon instead of silicon and massive computer memories made of organic polymers that are long strings of carbon-based molecules. By using a more flexible carbon basis for these machines instead of the more rigid silicon basis, we are copying the basic chemical structure of living systems. In fact there is even speculation about the development of self-reproducing organic chips. As a result, even if we never encounter intelligent life somewhere else in the universe, our notion of humanity will more than likely be challenged by the intelligent machines of our own making.

But what if we *do* encounter intelligent life? What if the Face on Mars is the product of an ancient civilization? The reaction which people have to the mere IDEA of a ''Face'' on Mars gives us an indication of how fundamentally such a discovery would shake our assumptions about ourselves.

The Potential Social Impact of the Mars Research

Much of the social and cultural impact of the controversy over these unusual landforms on Mars will be determined by the reaction of the two major revitalization movements which are underway in the United States and throughout much of the world.

The culture shock generated by the accelerating rate of new information and technology in the Western world has produced on of the most fascinating mass revival movements of modern history. Protestant evangelical Christianity in the United States and throughout most of the third world is leading a return to fundamental Biblical values based on the scriptures as the literal word of God. In third world countries, evangelical Christianity is making strong inroads in predominantly Catholic countries. A resurgence of fundamentalist revival movements in Islam is also readily apparent among the Shiites of Iran and Lebanon with consequences which have changed global politics.

This resurgence has been a catalyst for a wider social and political counter-revolution in the United States. Evangelical fundamentalist Christianity and its political counterpart, the New Right, has openly challenged secular humanism and scientific theories associated with it, expecially evolution. For those who share the fundamentalist evangelical belief system, there is a world of certain Truth as revealed by the word of God in the Bible. The acceptance of the notion of evolution is not congruent with their belief structure at Level 3 and Level 4.

Certain reductionist scientists who are also Christian fundamentalists try to bridge the two frames of reference with the notion of Scientific Creationism. The core of the controversy is the nature and purpose of Man — the feminine is downplayed here. The patriarchal dignity and destiny of Man is based on his special creation by God. Without this special creation Man is only a clever animal with no spiritual destiny or purpose. In an age of uncertainty and upheaval the evangelical Christian frame of reference gives priority in its belief system to the complete transcendence of God.

The belief systems of the non-fundamentalist institutional Christian churches such as the Episcopalians, Orthodox, and Roman Catholics, give priority to the immanence or presence of God in Creation and the union of God and humanity in the Incarnation of Jesus Christ. In these frames of reference humans are co-creators with God. The dignity and purpose of humanity derives not only from a direct of indirect creation by God, but also by the coming of Jesus as the definitive manifestation and revelation of God's love. As a result, these frames of reference have more tolerance for ideas such as evolution and can admit a wider degree of flexibility in social institutions. Consequently, social and technological change oriented toward the recognition of women, birth control, social and economic justice, and even a Christian Marxism or a theology of liberation, are at least options for discussion. These are all controversial issues for members of these churches, but the debate is about the content of Christian humanism, not whether there can be such a thing.

Nevertheless, even within the institutional Christian churches, the effect of these two opposing revitalization trends is evident. For the most part institutional Christian churches have adopted a conservative or ''fundamentalist'' approach to theological teachings about the nature of God and the obligations of humans but there is also a contrasting liberal emphasis on the protection of the environment and the achievement of social justice.

The social and cultural consequences of surprising or perplexing information from Space exploration, such as the ''Face'' on Mars, will be influenced more by people's religious beliefs than by their scientific education. To a great extent this has already been borne out by those involved in the inquiry, and perhaps more importantly, by those, whether religious or not, who dismissed the topic as not worthy of further inquiry. Contrary to the American cultural notion that facts are evaluated objectively and then accepted or rejected, there is every indication that our beliefs and values censor what we perceive and how we react to it.

This is perhaps the most important result of our armchair adventure to Mars. Science is a product of social and cultural systems. It is the result of a shared system of meaning and values which make up the everyday world of the inquiring person. From a strictly reductionist perspective, there is little need to spend more time on the question of the ''Face'' on Mars. However, from a systems perspective there is every reason to amplify the scope and intensity of the inquiry. For instance, the ''Face'' on Mars provides a ready-made opportunity for a joint United States-Soviet exploratory mission to Mars to create a common ground in space which cannot yet be found on Earth.

CHAPTER FOURTEEN

Earth to Mars

In an essay called ''From Versailles to Cybernetics,'' in his book *Steps to an Ecology of Mind,* Bateson traces the international violence of the twentieth century to the symmetrical structure of the treaty of Versailles which ended World War I. The pathological or self-destructive elements of the treaty led to World War II and by extension (in my opinion) to the current dance of death between the United States and the Soviet Union. Cybernetics, that is the study of information flow in systems, offers salvation from the pathological structuring of relationships, in Bateson's view. By studying the structure of relationships and making the proper adjustments in the structure of our thinking at Level 3 and Level 4 — changes which we generally describe in religious terms as conversion — it is possible to avoid the mutually assured destruction of the nuclear arms race.

A narrow, competitive ''race'' to Mars would only worsen the combative structure of United States and Soviet relations which endangers the survival of *Homo sapiens*. A broader, cooperative inquiry could move us all into an asymmetrical structure which could prevent a nuclear winter and usher in a political and social spring.

A definitive resolution of the scientific and human questions raised by the ''Face'' on Mars will require a return mission with adequate instrumentation. At present the Soviet Union is planning an unmanned mission called Phobos, which will consist of two spacecraft which will leave in July of 1988 and arrive in January of 1989. The primary purpose of the mission is to perform a chemical analysis of the inner moon of Mars, Phobos. The mission will also be able to do high resolution imaging for mapping Mars, and is scheduled to rendezvous with Mars' other moon, Deimos.

There has been some discussion of a United States unmanned mission in 1990, called the Mars Geological and Climatological Observer or Mars Observer. This mission, at present, is not being designed to carry any imaging equipment. There has been some discussion of the addition of imaging equipment in more recent months. However, this mission is apparently on the chopping block due to the spending cuts which will be triggered by the Gramm Rudman Hollings Balanced Budget Act, which requires phased-in reduction in United States Government spending to balance the Federal budget.

A cooperative venture into space is not unrealistic. Halley's comet, which the thirteenth-century Florentine painter Giotto depicted as the star of peace in his painting ''The Adoration of the Magi,'' is already providing an occasion for internaional cooperation. Brad Smith, the head of the Voyager imaging team, is working with the Soviet Union's team on the Vega mission, which is studying Halley's comet. This mission is a two spacecraft fleet named for its two primary objectives, Venus (VE) and Halley (GA). (There is no ''H'' in Russian; consequently, the ''G'' is used in the transliteration of Halley's name.) Venus and Halley also describe the route of the VEGA mission. The mission first headed for Venus and left 2 balloon observation posts and the ''buses'' — as they are called — looped around the Sun and made their rendezvous with Halley's comet in March of 1986. The Jet Propulsion Laboratory in Pasadena, California is assisting the VEGA mission by providing precision tracking of the space craft, using the United States' Deep Space Network of radio telescopes.

The United States' Deep Space Network, along with the Soviet Union, is also providing precision tracking for a European mission to Halley's comet. Appropriately, the mission is called Giotto. The Giotto mission came within 300 miles of the nucleus of Halley's comet and is equipped to take high resolution images and perform a wide variety of scientific experiments. Similar international cooperation was shown in the Japanese mission to Halley's comet, which is called Suisei. The Japanese mission was more limited in scope and primarily concentrated on ultraviolet scans of the ''coma'', or glowing cloud, which surrounds the nucleus of Halley's comet and blows away from the Sun to form one component of the comet's tail.

According to the Institute for Security and Cooperation in Outer Space, based in Washington, D.C., the recent 1985 Geneva summit talks between President Reagan and Secretary Gorbachev have produced a wide variety of possibilities for cooperative space exploration. There has been speculation that many of these efforts may be bargained away in an attempt to forestall the development of President Reagan's Strategic Defense Initiative (SDI) or ''Star Wars'' as it is called by its critics.

The ''Face'' on Mars provides us with a special opportunity to build on the international cooperation generated by the study of Halley's comet. Such international relationships enable us to find and build upon our common humanity. A new technology — video conferencing — which allows participants in different television studios to meet with each other electronically, has led to the creation of ''real time'' video conferences or ''space bridges'' in which the participants are all present at the same time. Larger audiences may observe the actual conference or view a television broadcast at a later date. (Space bridges are now beginning to incorporate telephone conferencing and/or computer conferencing. Slow scan video, a process by which video images are transmitted by telephone, is also becoming an element in the widening concept of the space bridge.) The ''Beyond War'' space bridge linked San Francisco and Moscow on December 13, 1984. A second ''Beyond War'' linked people all around the world on December 14, 1985 in another example of the human interdependence our species must foster in order to prevent its self-destruction.

In many respects the mid-twentieth century culture of the United States with its movies of invaders from space and space exploration adventure stories on television has revealed a partially conscious social preoccupation with the possibility of other intelligent life forms. Rollo May, the psychologist, in the opening of his book *Power and Innocence,* reiterates a common theme in the social and behavioral sciences, namely, that the creative persons in a society often anticipate in their art, poetry, or writing the challenges society will face. May likens the creative person to the parakeet taken down into the mine to warn the miners of gas. Like the parakeet, the creative person often does not survive. Unlike the parakeet, the creative person's sacrifice is not always taken as a warning.

Despite the multiplicity of popular cultural images of extraterrestrial intelligence, one of the most puzzling and unsettling images of life beyond Earth has not really been explored. The mass society created by the electronic media is well prepared for kindly or malevolent beings in a variety of colors with body types ranging from salamanders to energy fields. In itself this is not surprising. Humans in the twentieth century have seen themselves run the gamut from brutish sub-human behavior to unparalled human achievement in the arts and sciences. In reality these fictional outerspace creatures are caricatures of both the negative and positive facets of being human. Through science fiction and the growing folklore about UFOs, people around the world play with various partial and incomplete definitions of humanity.

The one image of extraterrestrial life which has yet to receive widespread attention is that of humanity itself. Richard Grossinger, in his book *The Night Sky,* in an evocative essay entitled ''Flying Saucers and Extraterrestrial Life,'' cogently summarizes our fear and preoccupation with extraterrestrials as a fear of ourselves. ''Perhaps *we* are the night, the sky, and the rest of it is fields of golden flowers and astral levity, on to eternity.''

For those willing to listen to the parakeet in the mine, the warning is clear. Our greatest threat comes not from the sky but from our unwillingness to face ourselves. The ''Face'' on Mars challenges us on a scientific level but more importantly, it challenges us on a human level. The scientific dimensions of the riddle of the Martian sphinx create a political and social opportunity to move toward a world in which we are less afraid of ourselves and others on our own planet.

A Project for a Friend

Our armchair adventure on InfoMedia's Notepad has shown that people in different places with different backgrounds can work together in the pursuit of knowledge. A larger international inquiry into the ''Face'' on Mars involving researchers from both the East and the West would be an extremely difficult and complicated task, but one that could be greatly facilitated by computer conferencing and eventually by some type of video linkup or spacebridge.

There is the possibility of a cooperative United States and Soviet Union unmanned probe to Mars which could arrive as early as January of 1989. However, this will require a lot of intensive activity in 1986 and 1987.

Of course if both nations wanted to send their own unmanned probes at the same time, the United States could divert the Gallileo probe. Originally, Gallileo had been scheduled to leave for Jupiter in May of 1986 but its departure has been postponed to June of 1987. With the right imaging equipment Gallileo could stop off at Mars on the way to Jupiter or perhaps even stay at Mars.

In fact, this would be a good job for Ren Breck. He's has enough time to recover from the Nepal Blindness Project. He has also gloated a little too much while I have been struggling with the "Face" on Mars. In fact this will be a great opportunity to use the "emergency" ruse Ren used to rope me into this in the first place. No, I won't place a call on those antiquated voice lines. From the comfort of my armchair of Martian anthropology, I'll log onto Notepad and throw down my gauntlet in challenge. Of course, I'll wait until Sunday afternoon. . .

APPENDIX

The Boulder Paper

PRELIMINARY REPORT OF THE INDEPENDENT MARS INVESTIGATION TEAM:

NEW EVIDENCE OF PRIOR HABITATION?

(C) 1984 by Richard C. Hoagland, Organizer
Independent Mars Investigation Team

August 12, 1984 (Revised December 12, 1984)

"...outer space is as much a territory of the mind as it is a physical concept."

Robert Vas Dias

INTRODUCTION:

Given the present state of human knowledge, any scientific discourse pertaining to the past, present or future existence of extraterrestrial life-forms must by definition be assumption-filled and value-charged. When that discourse is directed not abstractly at the cosmos at large, but toward anomalous surface features on a near-by planet, the danger of opinion overwhelming cautious and meticulous inquiry is greatly increased, if not inevitable.

To appropriately formulate and solve a "messy" (1) or "wicked" (2) problem whose data base is complex, ambiguous, and uncertain, methods of problem "management" (3) requisite to the task must be employed. Casually dismissing possibly significant "evidence" (4) as a trick of light and shadow or as the projective aspect of human pattern recognition is obviously inadequate and ultimately does a disservice to the community of planetary researchers and to science as a whole.

The following paper is presented not as hypothesis or proof of any claim, but simply as a report on a set of provocative data whose possible implications are profound, and which therefore is deserving of the planetary community's fullest attention.

HISTORY:

In 1976 the United States dispatched to Mars two robotic probes code-named "Project Viking." The mission consisted of four spacecraft for detailed investigations on the surface and from orbit (5). In July of that year, the initial Orbiter (designated the "A" spacecraft) secured a high-resolution frame of a region called "Cydonia" as part of an intensive mapping sequence. Initial analysis of this frame by the Viking Imaging Team revealed an intriguing curiosity: a mile-wide mesa (Fig. 1)

resembling a humanoid "head." A copy of the frame and an
enlargement of the "head" region were subsequently released to
members of the press by NASA's Jet Propulsion Laboratory (Fig.
2). Official comment at the time (August, 1976) dismissed all
scientific significance (6), terming the resemblance merely a
"trick of lighting." Recent comments by members of the planetary
community on the subject of "the Face" reveal no deeper interest
or analysis in the eight years since the image was secured by
Viking (7).

In 1979, DiPietro and Molenaar (8) rediscovered the image of
the controversial feature in the National Space Science Data
Center files. Subsequently, they embarked upon an independent
program (a) to locate additional Viking images and (b) to perform
image processing techniques on the original digital data used to
construct those images, in an effort to extract the maximum
existing data which would serve to better characterize the nature
of this enigmatic feature. They were successful in pursuit of
both of these objectives.

The original Viking frame (35A72), showing the feature at a
very low sun angle (10 degrees), was soon joined by a second
high-resolution image (70A13) taken at a significantly different
sun elevation angle (30 degrees) and azimuth from the original
(Fig. 3). A qualitative comparison of the most controversial
features of the original "face" (eye sockets, bilateral symmetry,
and continuation of apparent "hair" around to the right side of
the object, as viewed by an observer) confirmed their existence
on the second Viking frame. Whatever the geologic explanation
for these features, their persistence at the second sun angle
effectively eliminated "a trick of lighting" as an adequate
scientific explanation. Furthermore, detailed image processing
by DiPietro and Molenaar revealed additional features on both
frames that provided additional levels of confirmation regarding
the humanoid appearance of this anomalous object (8), such as the
presence of an "eyeball" and "pupil" within the lefthand eye
cavity (Fig. 4).

In addition to the controversial "face," DiPietro and
Molenaar discovered a second unusual object on frame 70A13 -- a
"pyramid-like object" (8) (Fig. 5). Located within a few
kilometers of the previous anomaly, the object was characterized
by four apparent sides, four flattened planes, and peculiar
"buttressing" at each apparent corner. The scale of this new
feature was similar to the dimensions of "the face": 0.6
kilometers for the short sides, and approximately 2.6 kilometers
for the long sides. Data taken from the spacecraft datablock on
70A13 indicated that the western "pyramidal plane" was oriented
parallel to the spin axis of the planet, i.e., north and south.

In 1983 Hoagland, upon reviewing both the images and the
available literature referencing these features within the
context of existing Mariner and Viking geological analyses, came
to the conclusion that their origin was still an open question
(9). Measurements of "the face," specifically its orientation
relative to "north" and "east," as well as measurements of

various proportions within the object itself, revealed a set of orientations and dimensions consistent with some prior planning for this feature (9) (Fig. 6). Additional measurements revealed a probable connection, in terms of orientation and location on the surface, relative to a separate set of objects positioned several kilometers west-south-west of the original "face" (Fig. 7). And after intensive investigation of these objects, which indicated they formed a "rectilinear complex" oriented so as to afford a "geometrically best view" of the controversial "face," Hoagland advanced the tentative hypothesis that "the face" and the "rectilinear complex" formed an intentional construction.

He further proposed that, by means of the specific geometrical alignment of the axis of this "unit" (NE/SW), he could discern a possible motive for this construction: to observe the Summer Solstice sunrise over "the face," viewed in profile (subtending an apparent angle of 6 degrees) from the location of the "complex" (9) (Fig. 8). Utilizing the most recent calculations describing the cyclic obliquity shifts of the spin axis (10), Hoagland estimated the last epoch when geometrical conditions created the appropriate alignment of this Summer Solstice sunrise, by making use of the relationship:

$$\text{Sin } S = \sin @ \ / \cos c \quad [1]$$

Where @ corresponds to the instantaneous obliquity of the planetary spin axis, relative to a line normal to the orbit plane; c is latitude; and S is the sunrise point, measured in degrees north from True East on the date of the Summer Solstice (11).

Calculation of the current Summer Solstice sunrise as seen from the center of the "complex" (40.9 degrees N. lat) revealed that sunrise occurs in the present epoch (obliquity 25 degrees) at least 6 degrees north of "the face." However, assuming the "construction" was created in a different epoch (not an unreasonable assumption, given the present inhospitable environmental conditions), Hoagland derived the obliquity of Mars which satisfied the alignment conditions specified before, by inverting the relation to:

$$\text{Sin } @ = \sin S \text{ X } \cos c \quad [2]$$

Inserting the appropriate parameters, he discovered that -- irrespective of any additional assumptions -- sunrise over "the face" on the solstice would occur (and had occurred) whenever the planetary obliquity reached 17 degrees (9). (It is interesting that 17 degrees has been identified (12) as the obliquity at which the average equatorial and polar insolation during the course of the Martian year become equal -- with consequent implications for global climatological conditions and atmospheric pressure levels.)

Furthermore, consultation with the table of changing obliquities over the past several million years furnished by Ward (10) allowed estimation of the epoch when the geometrical conditions described previously were last fulfilled: 0.5 million years BP (before present) (9).

Finally, in extending the "solstice sunrise sightline" northeast from the center of the "rectilinear complex" (or "city") across the "face", Hoagland noted an intriguing coincidence: the line of sight intersected a peculiar "cliff" lying crosswise on the ejecta blanket of a distant rampart crater, some 15km beyond the "face" itself. Most notable about this intersection was the fact that it took place precisely at the northwest end of this abnormal linear feature -- which lies at right angles to the flow pattern of the ejecta, and appears to be a raised rectangular object approximately 3 km long and several hundred meters wide. It appeared to be unique, insofar as there were no remotely similar features on adjoining crater blankets.

Measurements of this feature's orientation, relative to the meridian, revealed that it was tilted with respect to North by an amount identical (within the limits of measurement) to the centerline of the "face" and the major "structures" in the "city." Furthermore, another sightline -- from the center of the "city" past the "chin" of the "face"-- coincided with the southeasternmost limits of this "cliff" -- creating the rather remarkable coincidence that this feature, as seen from the center of the "city" some 25 kilometers away, subtended the precise angular limits of the "face" as viewed from the center of this "rectilinear complex."

Whether this relationship was due to chance, or in fact signified another aspect of a conscious design for the "rectilinear complex / face" relationship (such as a deliberately-created "artificial horizon" behind the "face" to mask out the relief of the background crater), Hoagland could not determine from the existing geometrical data. What was inarguable, however (within the current obliquity model), was the fact that half a million years ago, the Summer Solstice sunrise took place directly over the "face," and its distant strategicly-placed "cliff" -- whether anyone was there to witness it or not. Lasting a few thousand years (out of the approximately 120,000 years required to complete one obliquity cycle), the geometric alignment creating this Solstice sunrise spectacular would return approximately every million years -- the "amplitude envelope" required for Mars to tilt its full +/- 12 degrees around the current mean of 25 degrees (10). Thus, the Solstice relationship between the "rectilinear complex" and "the face" also obtained 1.5 million years BP; 2.5 million years BP; 3.5 million years BP, etc. No unique date, using the astronomical alignment postulated by Hoagland, was possible -- only a "minimum age" of 0.5 million years.

Hoagland did note, however, an interesting coincidence. Many workers (12, 13, 14) have commented regarding the anomalous polar regions of Mars. Composed of exotic "layered terrain," these regions are free of impact craters down to the resolution limits of the Viking Orbiters -- a "few hundred meters" (13) (Fig. 9).

Possible explanations for this lack of cratering over an area exceeding 10(5)km(2) include a very young age for the current icecaps of Mars -- on the order of several million years (12). The possibility that these recent geologic features on the planet formed as a result of a recent "favorable environmental event," allowing the design and placement of the anomalous objects at Cydonia, could not be totally discounted, according to Hoagland (9). Other anomalous geologic and climatalogical "fossils" could also indicate a possible "recent" change in the Martian environment (15).

After consultation with one of the investigative team members (Pozos), in which the obvious terrestrial astronomical, archaeological and anthropological assumptions were thoroughly discussed, it was decided that a multi-disciplinary "independent Mars investigation team" should probably be pulled together. Its purpose would be to review both the original work of DiPietro and Molenaar, as well as the new data and questions raised in the preceding paragraphs. Thus, in December 1983 the present ad hoc team was formed, consisting of members at a variety of institutions across the United States with backgrounds ranging from physics to archaeology, from geology to anthropology. The investigation proceeded via an on-line computer conference, furnished free-of-charge by InfoMedia Corporation, of San Bruno, CA., courtesy of Ren Breck.

This is a report of that preliminary investigation.

DATA:

Beginning in January 1984, the "Independent Mars Investigation Team" -- consisting of approximately a dozen members -- began a review of the preceding work (16). Arrangements were made for access to the digital data tapes of the Viking frames in question, mainly 35A72, 70A11 and 70A13, through the tape libraries at USGS, Flagstaff, AZ, and JPL. In addition, one of the team members (Brandenburg) began efforts to locate additional Viking images taken of the region, particularly those secured at "morning light" of the right hand aspect of "the face." A parallel effort, to examine selected regions of Mars in other Viking frames via the resources of the NSSDC, at Goddard, was initiated by three team members (Brandenburg, DiPietro, Dolphin). This "sweep" was designed to familiarize the team with other areas having geologic features similar to the Cydonia examples, as well as terrain types greatly dissimilar. All geologic maps prepared by USGS from Mariner and Viking missions to the planet were then ordered, on which the "old" and "new" Viking imagery was placed in context.

The following Viking frames have been secured on EDR digital tapes, for subsequent review and analysis by the "independent team":

35A72, 70A11, 70A13, 72A04, 43A01-04, 86A08, 86A10

In addition, a large number of negatives and prints of these and other regions of Mars were obtained from a variety of sources (NSSDC, USGS, JPL) for assistance in better characterizing the surface of the planet. Both rectilinear and orthographic versions (for alignment calculations) were used in the subsequent analyses. A request was made of Davies (17) for entry of the Cydonia region (35A72, 70A11 and 70A13) into "The 1982 Control Net of Mars," both for better positional information and more accurate orientations, relative to "north" and "east," of the unusual objects at Cydonia.

Finally, a library of relevant literature references was collected (Dolphin, Beatty), including all major works published by Carr et al. on the results of the Viking missions, particularly the view of the planet seen from orbit. A computerized literature search, through the Dialog computer service, was also performed to this end, in addition to the purchase of all major volumes on the subject.

ANALYSIS:

The USGS data tape of 70A13 (30 degree sun elevation angle) was selected for early "unpacking" and image reconstruction, in order to verify DiPietro and Molenaar's claim for "the existence of a second frame of 'the Face'" (8). The data was run on a PRIME 400 mainframe, with hardcopy (8X10 Polaroid) prints, negatives, and slides (35mm) available as records.

As can be seen (Fig. 10), the unprocessed Viking frame does indeed confirm the existence of the controversial "face" at the higher sun angle. Various contrast "stretches" bring out significant detail on the previously shadowed side (Fig. 11), allowing bilateral measurements and proportional analysis of the "base platform" as well as the putative "facial features." Channon, following Hoagland (9), performed the first proportional measurements for the team, confirming bisymmetry between "eye sockets," central "nose ridge," and apparent side edges. Variance of the "facial features" is 2%. Width measurement comparisons of the "base platform" edges reveals a greater variance, particularly between western and eastern sides (8%). The orientation of the "mouth" (90 degrees to the central "nose ridge") and extension west from the centerline to a point directly under "the eye" on that side, are in conformity with human proportions. A similar extension of the "mouth" east, into the more shadowed side (even at the sun angle of 70A13), is more difficult to delineate. However, a raised feature which lies at the edge of the "facial platform" at the end of the western extension of the "mouth" appears to have an exact counterpart at the eastern extension of the same line-like feature. A one-to-one comparison of the left and right bisymmetry, produced by "flipping" the left half of 35A72 over to create a "mirror image" and projecting this reversal onto a simultaneous projection of 70A13, reveals several points of correspondence -- including detail at the corner of the "eye," and detail under the "eye" (Fig. 12).

Finally, an isocontour map utilizing the gray scale information in the pixel array over "the face" was created, using a standard plotting program adapted to the outputs of the unpacked Viking tape of 70A13. The result (Fig. 13) from the PRIME provides additional information on photometric highlights and shadows, but does not contain any direct topographic information. Step interval was every 5 DN numbers over the 256 range.

Brandenburg's search for additional Viking frames over Cydonia produced four new images (673B54, 673B56, 753A33 and 753A34), the latter two at "morning light" with illumination from the direction opposite to those of 35A72, 70A13, 673B54 and 673B56. Photographic enlargement of the negatives of these new images (from NSSDC) provides additional confirmation regarding the "base platform symmetry" of "the face." Detail on "the face" itself, however, is poor -- due to the inherently lower image resolution (the result of being taken five times farther away than the two high-resolution frames -- 35A72 and 70A13). The presence of a righthand "eye socket" and a morphology for the lefthand "socket" (at lighting similar to that for the right "socket" in 70A13) is discernible, however, and provides an additional verification for the presence and general similarity of these features (Fig. 14).

Similar treatment was afforded the "pyramid" claimed by DiPietro and Molenaar to exist a few kilometers southwest of "the face" (8). Unprocessed PRIME images from the Viking data tape confirmed an object with the features described by DiPietro and Molenaar, with one exception: adjustments in the contrast "stretch" (to allow detail in the shadow to register on the same image as detail on the sunlit side) confirmed a morphology for this "unusual surface feature" originally suspected from the DiPietro and Molenaar data by Hoagland (9). The new images reveal a distinct five-sided appearance to the "D&M pyramid"; subsequent measurements indicate an object (Fig. 15) with three "short sides" of equal length (1.6 km), two "long sides" of almost equal length (2.6 km), and corner "buttresses" (five) indicated by appropriate shadowing. Detailed measurement of the two southern "buttresses" indicates a length discrepancy between the S.E. member and the S.W. member (Fig. 16). Assuming equal length originally, the results indicate a significant amount of debris accumulated around the S.E. corner of this object. Visual examination suggests a source for this debris: the collapse of a substantial part of the S.E. plane of the "pyramid," and subsequent flow around the S.E. corner.

Estimated depth of the material at this corner is in excess of 100 meters. The pattern of albedo variations in the vicinity of this corner is consistent with a "debris fan," with highlights and subtle shadings indicative of flow fronts distributed radially to the projected region of collapse.

Examination of the "pyramid" itself reveals two possible causes of this debris flow: a circular feature indicative of an impact crater, and a larger "domed uplift" feature (Fig. 17). The presumed crater lies on the lowest flank of the S.E.

pyramidal plane, while the "uplift" feature appears to have deformed the regularity of the pyramidal shape near the apex, and has bright and dark areas consistent with a raised "bubble" catching sunlight from the N.W. -- the lighting geometry in all three images of this feature -- 35A72, 70A11 and 70A13. The origin of the "dome" and its relationship (if any) to the "impact crater" is currently unknown. The sequence of events leading to the collapse of a major portion of this "pyramid" can only be inferred. But the energies known to be associated with impact features of this scale (100 meters) should allow a future

quantitative estimation of seismic effects to be expected from such an impact, and the likelihood of a triggered collapse in various materials of differing cohesion.

Finally, in response to a suggestion from Channon, initial artist for the team, a superposition was constructed: an image of the "D&M" underneath a transparency of another five-sided figure -- Da Vinci's famed "Man is the Measure of All Things" -- outlined in a circle (18). With the scales properly adjusted, the proportions of the classic human figure -- hands outstretched -- over the proportions (reconstructed) of the "D&M pyramid" are the same (Fig. 18). The significance of this coincidence is open to interpretation.

Because of the complexity, and the unavailability of adequate computer time, the "rectilinear complex" ("the city") noted by Hoagland W.S.W. of "the face" (9) did not receive equal study or analysis, compared to the preceding "unusual features." The existence of a set of objects -- some of them pyramidal in form -- oriented parallel to the central axis line of "the face" several kilometers away was verified (Fig. 19). The internal rectilinearity of this array was also verified, via a superimposed grid which allows alignment of a number of interesting features and apparent straight edges. Components at precise 90-degree angles to an axis oriented through "the face" are also apparent, after superimposition of this grid (Fig. 20). A cursory examination of a regional geologic map for the predominant fault patterns in the area (19) does not appear to support this as a primary source of the alignments, but further work is indicated.

One highly controversial feature of "the city," identified by Hoagland as "the honeycomb" (9), remains to be adequately verified and explained (Fig. 21). Appearing as a "lattice-work array" of both regular and irregular crosshatchings in approximately a 1-kilometer-square region of a processed version of 35A72 within "the city," this feature's reality as a true latticework has been severely questioned. While dismissed as a "processing artifact" by some members of the team (Cordell, DiPietro), the suspicious cross-hatching of the surface relief, as determined by the shadow from an adjacent pyramid-shaped object falling over this area, is consistent with a three-dimensional structure on the surface in this region. The inferred relief of the "honeycomb area" determined from this shadow geometry is also a pyramidal shape, with a slope angle of

approximately 15 degrees. The "upper deck" of this apparent
pyramidal "honeycombed" object lies on the exact central axis of
the shadow-determined pyramidal object, and is symmetrically
adjacent to another feature located N.E. The "contact" between
these two areas is sharp and nonrespective of the obvious relief
of the N.E. object. Furthermore, a sharp albedo change marks the
boundary between "the honeycomb" and the object N.E., termed by
Hoagland "the fort" (9). Even on unprocessed Viking images, this
contact is apparent, and its linear aspect over a distance
exceeding 2 kilometers -- despite significant relief along this
contact -- is puzzling (Fig. 22). Hoagland's observation, that
the contact "reflects a sharp transition between underlying parts
of 'the fort' and an overlying 'honeycomb' which veils the
supporting relief..." seems consistent with an unprocessed image
taken by Viking at a higher sun angle. On frame 70A11, the
relief previously terminated by the apparent "veiling" contact
can be seen to extend faintly toward the southwest -- as if seen
dimly through a suspended layer of material between the ground
morphology and the Viking cameras 1500 kilometers above (Fig.
23). The lack of obvious geologic explanations for this
behavior, and the implications inherent in verification of such a
"honeycomb structure," make this problem deserving of additional
attention.

Measurements conducted on an orthographically rectified
version of 35A72 raise additional questions to be answered. If
lines are drawn between the "D&M," "the face," and the eastern
object in Hoagland's "city complex," then connected back to the
origin at the "D&M," the result is a right triangle. The
"straight wall" of the eastern object ("the fort") narrowly
defines the connection of this line with the "D&M." The placement of the vertex on "the face" is subject to more error
depending on which "feature"is selected: "eyes," "nose," or
"mouth." The best fit seems to derive from placement between the
"eyes," the final "sightline" chosen by Hoagland for his solstice
alignment calculations (9). In another interesting
determination, a "mirror image" of this triangle placed congruent
with the line connecting "the fort" and "D&M" places the "mirror
image vertex" on a point on "the city's" western edge. This new
isosceles triangle (first noted by Rautenberg) connects the three
central features ("city"/"face"/"D&M") cited by Hoagland (9) as
forming the central "Cydonia unit" (Fig. 24). The need for
additional study of these possible relationships is indicated.

The suggestion by Hoagland, that the apparent alignment
between the "city" and "face" formed by the base of the
previously mentioned isosceles triangle also forms a "Solstice
Sunrise Alignment," was also investigated (Fig. 25). New
positional data and orientation of the key objects ("the face"
and "the city"), courtesy of Davies (17), allowed recomputation
of the last epoch when Mars' changing obliquity would produce the
required solstice alignment between "the city" and "the face,"
through equation 2. The new orientations derived by Davies for
"the face" and "city" differ by 4.5 degrees (28 versus 23.5) from
spacecraft data used by Hoagland in the original alignment

calculations (9). The resulting revised epoch from Ward (20) for solstice sunrise along a line passing through the "eyes" from the center of "the city" is little changed from Hoagland's original estimates: 400,000 years. Considering the uncertainties in the obliquity calculations, spacecraft pointing and navigation, etc., this result confirms the fact that several hundred thousand years ago solstice sunrise did indeed occur behind "the face" -- as seen from the center of the "rectilinear complex." Whether such geometry was deliberately designed is, of course, a different question.

The extension of the 1982 Control Net of Mars (17) to Cydonia, as a result of this investigation, has two additional results: the "northern alignment" of the western plane of the "D&M pyramid" originally claimed by DiPietro and Molenaar (8), is somewhat in error (by 4.5 degrees); but one controversial measurement originally conducted on "the face" by Hoagland (9) has been confirmed.

A north/south line drawn diagonally, crossing the central axis of symmetry between the "eyes," intersects the W.S.W. corner of the "mouth" -- at an angle equal to the tilt of the entire "face" off north/south: 28 degrees (Fig. 26). A line run parallel to the central axis line, beginning at this S.W. "mouth" intersection point, passes directly through the "peak" in the "eyebrow" of "the face." It also passes over the "pupil" claimed to exist in the "eye socket" by DiPietro and Molenaar (8). The angle of this parallel line from the north/south diagonal is also 28 degrees: N.N.W. A "mirror image" of the diagonal line, drawn over the shadowed right hand side, also crosses the (estimated) N.E. corner of the "mouth" and under (in terms of the central axis of "the face") the corresponding right hand "eyebrow" peak. The apparent existence of so many symmetrical proportions, and their apparent connection with the overall orientation of this object, continues to make this a most "unusual Martian surface feature."

In the course of examining several hundred Viking images, as part of an effort to characterize these "unusual features" within the broader Martian context, team members have identified several additional anomalies which seem to fall within the category as defined within this paper. These range from an array of rectilinear "grooves" or "tunnels" appearing in the ejecta blanket of a pedestal crater, to an anomalous object -- sharp-edged and pyramidal in shape -- perched on the edge of an eroded impact feature (Fig. 27). Other unusual objects the team thinks deserve closer attention include: a peculiar "gate pyramid," consisting of a pyramidal sharp-edged object flanked by two parallel "walls" (Fig. 28); a regular pentagonal-shaped object, with a hexagonal "courtyard" perfectly centered within; a linear rectangular feature extending approximately 4 kilometers, with regular "bumps" spaced along its length, located a few kilometers from a mesa with an array of unusual surface "objects" displayed in an apparently non-random pattern (Fig. 29).

Very preliminary measurements of these noted objects, both in terms of location on the surface of the planet and orientation, reveal some interesting patterns. All the objects described lie along an "arched crescent" stretching from Cydonia halfway around Mars to the Elysium plateau. All (with the exception of the linear object with the "bumps") lie near the zero kilometer reference datum, as published in the Atlas of Mars (Fig.30), The 1:5,000,000 Map Series (21). The exception -- the linear feature with the "bumps" -- lies at the 4 kilometer elevation on the slope of Hecates Tholus.

In terms of specific orientations, these objects are even more interesting. Beginning with the "tunnels," the datablock on the Viking frames indicates an exact (within the errors of the datablock) north/south east/west orientation for this rectilinear array of sharp features apparently incised in the surface of the ejecta blanket of the associated crater. A few kilometers away, the equally enigmatic "crater pyramid" -- a massive rectangular object with dimensions similar to those of the "D&M" -- lies at a provocative 45-degree angle to the cardinal compass points. Further interest in this object was generated among the team when it was discovered that it lies adjacent to the long-discussed "furrowed ground" originally noted by the Viking Imaging Team, and that it is the highest point for more than 100 kilometers in all directions. The "pyramid" also lies almost exactly halfway between the Martian equator and the north pole -- at 46.3 degrees North, 353 West. The best frame is 43A01.

The so-called "gate pyramid" and its associated unusual features (the "pentagon" and an array of lesser anomalies) lie within 200 kilometers of the original Cydonia examples (specifically, S.W.) Special interest was generated over this feature because the "gates" appear to front on an ancient plain, possibly once containing water, with the "pyramid" standing at the entrance to an interior "lagoon" appearing as an interior "basin" in an elongated "island." Less than 20 kilometers away, the peculiar "pentagon" lies amid an array of puzzling features, including a suggestion of more "honeycomb." This frame (72A04), centered at 38 N., 12.6 W., deserves additional processing, in the opinion of the team.

The long, narrow, rectangular 4-kilometer object lying on the flank of Hecates Tholus appears to be a very different kind of anomalous object. Termed "the runway" to allow easy identification via the computer conference (which lacked on-line graphics capability, and thus encouraged easy-to-identify nicknames for objects of potential interest), this feature is all the more puzzling for its substantial relief (exceeding 300 meters), regular "bumps" (spaced approximately every 300 meters), and, most importantly, its orientation: exactly east/west (again, within the errors of the frame datablocks). Equally intriguing is the array of objects on the adjacent mesa, which are unique to this one mesa (other flat-topped mesas in the immediate vicinity are absolutely bare of similar relief). This, the pattern of these objects, and their apparent relationship to the linear "runway" with the "bumps," makes this frame (86A08) another candidate for processing, in the opinion of the team. The location of the center of the frame is: 34.6 N., 213 W.

DISCUSSION:

Much scientific progress has been made historically by establishing a pattern for anomalous phenomena before attempts at a theoretical explanation (22). Within the constraints of time and resources available to the Independent Mars Investigation Team, such "pattern recognition" was attempted for the limited data set uncovered by the team.

Hoagland initially called attention to the possibility that the Cydonia "site" was on the shores of an ancient lake, or bay (9). Following this suggestion, several team members (Beatty, Brandenburg, Dolphin, and Pozos) accepted this as a working hypothesis, for reasons ranging from the purely geological to anthropological. Immediately north of the "rectilinear complex" and "the face," the surface of the Cydonia plain appears patterned with a series of polygonal "cracks" or fractures. One explanation for these unusual features is permafrost cracking, though the scale is significantly larger than in terrestrial analogs (23) and implies a considerable depth to the frozen sediments, or long thermal cycling possibly related to the obliquity shifts of Mars. The presence of these cracks, then, is consistent with the idea that water at some time lay in the lower areas of the Cydonia region. The anthropological reasons for finding this hypothesis attractive stem from the well known relationship on Earth between "settlements" and water, and the presumption that a similar need might exist elsewhere.

A continuation of this reasoning soon led to the discovery that all the northern plains north of the Cydonia frames lie below the zero reference datum (21). This in turn produced the suggestion that perhaps the northern plains -- sparsely cratered, lying in some regions several kilometers below this reference datum, and patterned over the "deepest" regions by larger versions of the polygonal "cracks" -- was the sedimentary deposit of a former "ancient ocean," which would have occupied essentially the entire northern hemisphere of Mars. Other workers have come to similar conclusions (24). It was this intriguing suspicion which then led to the discovery that all the anomalous objects identified by the independent team, and described in the preceding pages, lie along the "shoreline" of this hemispheric basin.

If the objects cited formed concurrently with the existence of this proposed "ocean," then the erosion expected along any "continental/oceanic boundary" could provide a possible natural explanation for the terrain anomalies noted previously (25, 26, 27) between "the high-latitude northern plains" and the "old cratered terrain." This in turn could have provided, along this natural "continental/oceanic boundary," abundant natural "morphs" for deliberate reshaping. In the opinion of the team, a previous "warm, wet epoch" in Mars' early history (12, 28, 29) could have supplied sedimentary deposits to fill these lowland plains, by erosion of the ancient cratered highlands to the south. In some respects, Mars resembles a "Victorian model of the earth," with fixed topographic "highs" and "lows" corresponding to classic (pre-plate tectonic) ideas regarding elevated continents and depressed ocean basins on Earth. Similar ideas have been expressed by Nummedal (30).

Numerous dome-like formations on the northern plains are thought by many workers, including Allen (31) and Hodges et al. (32), to be analogous to terrestrial landforms found in Iceland, notably "subglacial volcanos." This is also consistent with the possible existence of a former ocean, as water (either in liquid form or as ice, which subsequently melts) is the chief agent for producing the recognized morphology of these "pseudocraters." If the water contact occurred via ice (possibly when such an ocean was frozen, after the Martian climate deteriorated drastically), Allen estimates its thickness as between 100 and 1200 meters, from the current elevations of the topographic forms. Hodges et al. go even further, postulating "ice caps," for selected regions of the plains, of "up to several kilometers" (32), consistent with a frozen ocean.

Supporting this hypothesis (of an early ocean), one of the team members (Brandenburg) has marshalled a variety of existing geochemical evidence. Toulmin et al. (33) have summarized efforts to translate elemental abundances from the Viking Lander Inorganic Chemical analysis into expected oxides, then into expected mixtures of minerals which might satisfy the chemical data. Best matches have been found for altered products of basaltic rocks, dominated by smectite clays of nontronite, montmorillonite and saponite (33). Several problems for the proposal that the surface materials consist of clays have arisen, however, notably their formation in the absence of (current) liquid water, discussed by Gooding (34), and the apparently contradictory evidence of red (oxidized) surface materials in the presence of clays requiring a reducing environment in which to form, noted by Baird et al. (35).

Brandenburg observes that the presence of both a former oceanic environment for Mars and an atmosphere containing substantial amounts of free oxygen could conceivably remove these apparent contradictions.

The substantial amounts of heavily-oxidized iron-rich "fines" detected by the Viking Landers, and the indirect argument for the presence of smectite clays, could be explained as a mixture: the oxides form through exposure to free oxygen on the "land"; the clays form through chemical alteration of original basalts in the presence of abundant water beneath "the ocean." The epochs represented by these two materials need not be identical; the "free oxygen epoch" could conceivably come some time after the early "oceanic period." The fact that Viking landed on the proposed "floors" of the ocean, and thus sampled (presumably) sedimentary material in situ, is consistent with the discovery of enrichments to the samples of certain salts (chlorine), as well as the presence of "duracrust" possibly formed by the upward migration of water and evaporation at the surface of the ancient "ocean floor." Brandenburg notes that nontronite and montmorillonite clays on Earth are evidence of a "mature ocean," appearing only during the Mesozoic Era as a result of changes in terrestrial ocean chemistry (36).

The point is that to even conceive of the unusual surface features identified by this team as being anything more than the products of exotic geological conditions requires the present inhospitable Martian environment to have been different some time in the past, possibly the "recent" past -- a few million years. Curiously, in the existing geochemical data supplied by Viking, such a possibility is not excluded.

While previous laboratory simulations of the current Martian atmospheric environment and ambient conditions (especially the availability of strong ultraviolet light) have sought to explain the presence of the Martian oxides as the result of UV photostimulated oxidation over geologic timescales (37), more recent work (38) does not support this mode of oxidation for the surface material on Mars. A return to a previously suggested origin for these oxidized materials, first proposed by Sagan et al. in 1965, may be favored (39): that the presence of abundant ferric oxide, mixed with the silicates being blown around the Red Planet, has its likely origin in a previous atmospheric epoch in which abundant free oxygen was present. We view this as an elegant solution to the previous geochemical contradictions, but also point out that to adopt such a hypothesis raises profound new questions. On Earth, the first extensive oxidation of the surface has been traditionally ascribed to the appearance of living organisms, and the liberation of free oxygen by photosynthesis (40). In terms of the presently understood conditions and timescales for the "early, wet epoch" for the Martian environment (12), and the rates of biological evolution -- particularly its dependence on average ambient temperature (41) -- it seems difficult to accept an indigenous origin for equivalent Martian microorganisms capable of altering the primordial atmosphere from a reducing environment to an oxidizing one, consistent with the extensive surface oxidation. Possible "panspermic" models for the arrival of such proposed organisms on Mars may eventually have to be considered.

Finally, in terms of "recent" geochemical anomalies, the results of the Viking Labeled Release experiment may be relevant to this discussion.

In the final NASA report by the Principal Investigator for the Labeled Release Experiment (Levin, of Biospherics, Inc.) (42), the authors term their own efforts to reproduce their Mars data, using non-biological chemical reactions, "unconvincing on critical points." The main obstacle, they state, is the instability of the proposed chemical "super oxide" in the soil required to account for the results of their experiment (in the non-biological model). Hydrogen peroxide breaks down rapidly under ultraviolet light, raising questions regarding both its current formation in the Martian soil and its persistence across geologic timescales. Without discarding a chemical explanation for the Mars results of this instrument, it is possible to come to a somewhat different conclusion: that the instability of peroxides on the present Mars is an indication of a recent alteration of the Martian climate -- from an epoch when such peroxides could form (in an oxygen-rich environment, under very dry and cold conditions) to the present environment, which lacks the key ingredient for the above scenario, free oxygen. If this tentative suggestion has validity, it could explain both the puzzling presence of abundant oxides in the current Martian soil, and their persistence in the face of conditions which are systematically destroying them -- as a "relic" of a "recent" drastic change in the Martian environment.

Other evidence in favor of such a "discontinuity" comes from climatalogical inferences possible from the Viking Orbiter data.

The presence of the remnant polar caps -- one composed totally of water ice and the other apparently solid carbon dioxide -- remains one of the outstanding anomalies of Mars (43). Current efforts to explain this hemispherical asymmetry reach beyond purely radiative models, and draw upon the fact that the major dust storms on Mars currently take place in the southern hemisphere. This (according to the model) could screen the southern polar cap from the full effect of sunlight during the otherwise warmer southern summers. (This hemispherical asymmetry in theoretical temperatures is due the fact that Mars is currently closer to the sun when the southern hemisphere is tilted toward it.) The "dust screen" thus produced in summer could protect solid carbon dioxide from total sublimation, despite theoretically warmer southern temperatures. This, in turn, could result in a current accumulation of more solid carbon dioxide on the remnant southern cap, according to the model.

An alternative, less ad-hoc, explanation is that the current asymmetry is but another "relic" of a recent drastic change in the overall climate of the planet. If temperatures rose enough to allow melting of the currently extensive permafrost reservoirs beneath the northern plains (44), subsequent condensation of this water on the "cold trap" provided by the raised landmass beneath the current northern polar cap could easily produce a polar icefield with a water composition. With most of the available water on the planet lying in the northern hemisphere, and the high, arid "desert" surrounding the south pole providing initial temperature asymmetries, the possibility that a subsequent return to glacial conditions could lead to a predominantly frozen carbon dioxide cap in the southern hemisphere, and a water ice cap in the center of the former "polar ocean" girdling the north, cannot be excluded.

Additional evidence consistent with this idea comes from the structure of the caps themselves. The remarkable "layered terrain" apparently making up both caps (regardless of their different compositions), and their equally apparent youth (as derived from an absence of discernable craters on either cap), could indicate a recent climatological "event" leading to their presence. Visible unconformities within the layers, particularly those in the northern cap, indicate a sharp change -- from an epoch of deposition of the layers to an erosional epoch, then a renewed epoch of deposition. The data indicate that this "unconformity" occurs at least twice within the otherwise remarkably consistent layering (45). Thus, in the "youngest geologic features on the planet" (12) we see evidence of abrupt changes in whatever mechanism led to the formation of the current polar caps. Via cratering curves, the approximate epoch when this proposed change might have taken place can also be calculated (45): possibly as recent as a few million years ago.

Other supportive evidence in favor of a "recent" change in paleoclimatic conditions for Mars comes from analysis of the circumpolar Great Sand Sea, and the presence of wind-eroded ventifacts.

The Great Sand Sea is a massive field of sand dunes (over a million square kilometers) almost completely encircling the northern polar regions. Its presence has been cited as evidence of some kind of intimate connection or causal relationship with the polar cap itself (45, 46). Tsoar et al. state that "The dunes seem to be currently active . . . (and) the lack of well-developed longitudinal dunes implies that the dune field is young." The authors do not give an estimated age, but other analysis of sand dunes on Mars and the erg (sand sea) around the pole, by Breed et al. (47), produces estimates of "20 million years for the formation of a mature erg." These data and inferences from terrestrial analogs are consistent in the opinion of some members of the team, with a sedimentary origin for the

material within the dunes (from the exposed "floor" of the
proposed ancient northern hemispheric ocean) and the formation of
both the water ice cap and the associated erg as a result of a
"recent" drastic change in climate -- with all the attendant
implications.

Other evidence supporting changes in atmospheric density,
wind velocities, composition, and ability to pick up suspended
dust and sand, comes from analysis of yardangs (eroded
streamlined forms) imaged by the orbiters and analyzed by Wesley
Ward (48). On the basis of geologic units in which yardangs
currently appear on Mars, Ward raises questions regarding the
possibilities for a "recent" epoch of severe wind erosion on the
planet. This implies a significantly higher atmospheric density
than in the current Martian environment. El-Baz et al. (49)
point to the discovery of pitted and fluted rocks, found in
Egypt's Western Desert, and compare their surface characteristics
with those of similar rocks seen in the pictures relayed from the
Viking Lander cameras, rocks previously thought to be vesicular
basalts (with the pits formed by escaping gas bubbles in former
lava). The new data indicate that the morphology of the rocks
seen by the lander cameras can be created solely by wind erosion,
with the conclusion that "the Martian surface may be far more
wind eroded than previously thought" (49). The implications, in
terms of changes in Martian climate required to support these
erosive features, are apparent.

One cannot leave the subject of wind erosion without
reference to the notorious "pyramids of Elysium," and a
comparison with the "pyramids" cited in this paper in the
previously unexplored Cydonia region of the planet. It is
unfortunate that the "pyramids of Elysium," as examples of
"pyramids" on Mars, have received so much attention (50).
Although they have been explained as examples of wind-faceting
and tectonic processes (51), the application of similar processes
to explain the unusual objects at Cydonia -- in particular, the
remarkable "D&M" pyramid -- does a disservice to the Cydonia
objects -- truly "morphs of a different color." In the opinion of
the Independent Mars Investigation Team, the five-sided pyramidal
object at Cydonia (and a similar feature -- the "crater pyramid")
are of a completely different class than the too-often-cited
objects in Elysium. Whatever the eventual explanation for the
"D&M," it is the opinion of the team that superficial comparisons
with the "pyramids of Elysium" will not serve to increase our
understanding of whatever formed the "D&M," "the city," or the
most remarkable anomaly of all, "the face" itself.

Finally, a long-accepted indicator of a potentially different
Martian climate -- albeit a very ancient climate -- has been the
"runoff channels."

These arrays of apparently eroded river valleys have been cited often (15, 23, 26) as evidence for a former "warm, wet epoch" in the history of Mars. But because of the observed preference of these features for the ancient cratered highlands, many investigators have almost automatically assumed that the creation of the channels (from erosive processes extant during a brief denser atmospheric epoch concurrent with the bombardment which produced the cratered highlands) confirms the presumption that the runoff channels formed very early in the planet's history. Their existence, as features restricted primarily to the oldest terrains on Mars, has been seen as persuasive evidence that the "warm, wet epoch" was a concurrent byproduct of this brief formation period -- and disappeared soon after.

Pieri (52) has compared the morphology of the highly provocative runoff channels with their terrestrial counterparts, and has discovered subtle differences from valleys formed primarily by running water. The main difference, according to Pieri, is a tendency for the Martian runoff channels to begin in stubby, truncated tributary networks; this feature, Pieri notes (following Sharp, 1973), is more indicative of "sapping" -- underground collapse due to subsurface water or ice -- than to surface erosion due to rainfall allowed by a denser, wetter atmospheric epoch.

Pieri's second comparative point is aimed at what he terms "the immature nature of the Martian drainage system." He argues that the apparent simplicity of most of the runoff channels (in contrast to most terrestrial river tributary systems) is consistent with, at most, a brief epoch for such a presumed "warm, wet epoch" -- and one restricted to the earliest first few million years of Mars.

An apparent contradiction of this last point, however, has been noted by Carr (53), when he states, "This conclusion of poorly established drainage patterns and relative immaturity is somewhat at odds with the wide range in preservation of the small runoff channels, which suggests a range of ages, some of which must be quite old with respect to the present landscape."

Theoretical estimates of the longevity of a denser Martian atmosphere, from Pollack et al., argue persuasively that such an atmospheric epoch should have persisted for "at least several hundred million years." This would support the evidence of very old (and degraded) runoff channels seen underneath a system of fresher, newer ones, yet contradicts Pieri's assertions regarding the network's overall "geomorphological immaturity."

One way around this apparent contradiction, we suggest, would be the invocation of <u>two</u> <u>heavily</u> <u>erosive</u> <u>epochs</u> -- separated by several billion years.

The initial runoff channels would then be formed during the initial "heavy bombardment period"; the younger, crisper, and "immature" channels (to use Pieri's phrase) could be explained by a much more recent -- and briefer -- atmospheric "spike," consistent with the other anomalous climatological evidence presented in this paper.

It should be pointed out that the obliteration of craters in the ancient cratered highlands, by a severe erosive epoch, has been generally, but not exclusively, connected by most observers with the massive cratering events associated with the earliest era of the planet. A dissenting point of view, namely, that there is nothing inherent in the statistics of crater distribution and frequency to preclude a much more recent epoch of "crater obliteration" from a denser atmosphere, has been advocated by several investigators, most notably by Jones and Chapman (54).

Chapman and Jones' arguments for an erosional "spike" subsequent to the decline in cratering rates fits well with the pattern presented here: that a very late "event" somehow created a brief, denser and more temperate atmosphere, possibly due to other than strictly natural causes. In this model, the presence of "immature runoff drainage networks" preferentially incised in the oldest, most pulverized (thus most erodable) cratered highlands, is expected, as is the fact that such a proposed "late erosive epoch" would have been severely truncated in time -- consistent with the formation of Pieri's "immature" new channels.

The most severe constraint on these ideas, namely, the rate of cratering events subsequent to this "late erosive epoch," may be lifted in light of the recent "Nemesis" ideas proposed by Alvarez and others (55). An adjustment of the Martian crater dating curve, reflective of a periodic shower of fresh comets across the inner solar system triggered by the passage of the proposed "Nemesis star" through the inner Oort cometary cloud, or other non-uniform cratering events, could adjust the current estimates of how often new impact craters form on Mars, to replace those wiped out in the last erosive "spike" which we propose. This revision in the crater curve should lead to a downward revision in the estimated ages of all geologic units dated by cratering statistics.

CONCLUSION:

Based on this data, and on the analysis of the possible relationship of a number of anomalies in the existing literature regarding Mars, it is the opinion of the Independent Mars Investigation Team that the problems raised by this preliminary effort warrant the allocation of substantial additional resources.

Testifying to this conclusion is the recent decision by the International Program in Applied Systems Design at the Center for Research in Management, University of California, Berkeley, to assemble a multi-institution, interdisciplinary research group, in order to produce a full technical report on this material. (Members of the planetary community interested in possible involvement should contact Thomas D. Rautenberg, Administrative Director.)

In the light of certain implications, should study of this material conclude that these are the remains of a former inhabited epoch for the planet Mars, the independent team would encourage planning for an early Mars Return mission, to secure higher resolution imagery and complementary data. With lead times for unmanned missions approaching a decade, and with existing plans for the proposed Mars Geoscience Climatology Orbiter lacking in suitable imaging instrumentation (56), the concept of an alternative means of securing the necessary data seems prudent.

One possibility is a Galileo II mission to Mars.

Published costs and timelines for assembling a duplicate Galileo Orbiter (57), and flying it to to a planetary rendezvous, reveal that assembly of a second spacecraft from "spares" is an inexpensive, as well as a timely, means of obtaining potentially crucial imaging information on these anomalies within a reasonable time frame. Launch opportunities occur every two years (58), affording flexibility to mission planners commensurate with the flexibility inherent in not having to design a completely new spacecraft. Furthermore, studies of the applicability of the Galileo science instrumentation to Martian imaging and geochemical requirements, performed as part of a Mars encounter during a gravity-assist trajectory to Jupiter for the original Galileo mission (59), indicate no major problems for such a concept. Though not optimal, the current Galileo instrument design would allow a significant improvement in the quality of data on these unusual surface features. The application of other science instruments planned for such a duplicate spacecraft (the Near Infrared Measurement System and a proposed side-looking radar mapper), within the projected costs of $150 million for the mission, could add valuable basic information for efforts to place these anomalies within a broader geologic context.

With other nations (60) planning a return to Mars at the earliest opportunity (now less than two years away), it seems appropriate that the United States maintain leadership in this potentially important area of study. The "Galileo option" affords a cost-effective and timely means of maintaining that leadership, while furthering our scientific understanding of the problems raised in this preliminary study.

End

NOTES

(1)   Russel L. Ackoff, <u>Redesigning the Future</u>, Wiley (1979)

(2)   Rittel, Horst, and Webber, "Dilemmas in a General Theory of
      Planning," Policy Sciences 4:155-169 (1973)

(3)   C. West Churchman, <u>The Design of Inquiring
      Systems</u>, Basic Books (1971)

(4)   I. Mitroff, R. Mason, and V. Barbabba, "Policy as Argument:
      A Logic for Ill-structured Decision Problems," Management
      Science, 28:1391-1404 (1983)

(5)   Carl Sagan, "Viking to Mars: the Mission Strategy," Sky and
      Telescope, Vol. 50, No. 1, July 1975

(6)   David Chandler, <u>Life On Mars</u> , E. P. Dutton (1979)

(7)   Editorial reply, <u>The Planetary Report</u>, June 1984

(8)   Vincent DiPietro and Greg Molenaar, <u>Unusual Martian
      Surface Features</u>, Third Edition (1982)

(9)   Richard C. Hoagland, <u>The Monuments of Mars: a
      City on the Edge of Forever</u>, Prentice-Hall (in
      preparation)

(10) William R. Ward (private communication)

(11) Owen Gingerich, "The Basic Astronomy of Stonehenge,"
      <u>Astronomy of the Ancients</u>, The MIT Press (1979)

(12) William R. Ward, "Climatic Variations on Mars: Astronomical
      Theory of Insolation," J. Geophys. Res. 79:3375-3395 (1974)

(13) Bruce C. Murray and Michael C. Malin, "Polar Volatiles on
      Mars - Theory versus Observation," Science 182:437-443
      (1973)

(14) James A. Cutts et al., "North Polar Region of Mars: Imaging
      Results from Viking 2," Science 194:1329-1337 (1976)

(15) Harold Masursky et al., "Classification and Time of
      Formation of Martian Channels based on Viking Data," J.
      Geophys. Res. 82:4016-4038 (1977)

(16) Independent Mars Investigation Team: Hoagland, R. C.,
      Beatty, W., Brandenburg, J. E., Cordell, G., DiPietro, V.,
      Dolphin Jr., L., Pozos, R., Rautenberg, T.

(17) Merton E. Davis and Frank Y. Katayama, "The 1982 Control
      Network of Mars," J. Geophys. Res. 88:7503-7504 (1983)

(18) Da Vinci - Notebooks

(19) Controlled Photomosaic of the Mare Acidalium Southeast Quadrangle of Mars, Miscellaneous Investigation Series, U.S. Geological Survey (1981)

(20) William R. Ward, "Large-scale Obliquity Variations on Mars," Science 181:260-262 (1973)

(21) Batson, Bridges and Inge, Atlas of Mars: The 1:5,000,000 Map Series, NASA SP-438 (1979)

(22) Bateson, Steps to an Ecology of Mind, Bantam Books (1974)

(23) Michael H. Carr et al., "Preliminary Results from the Viking Orbiter Imaging Experiment," Science 193:766-776 (1976)

(24) James E. Oberg, Chapter 8, "Mars - A Closer Look," pp. 191-193 in New Earths, Stackpole Books (1981)

(25) Michael H. Carr, Viking Orbiter Views of Mars, NASA SP-441 (1980)

(26) L. A. Soderblom et al., "Martian Planetwide Crater Distributions: Implications for Geologic History and Surface Processes," Icarus 22:239-263 (1974)

(27) George A. Brook, "The Origins of Martian Fretted and Troughed Terrain," Third International Colloquium on Mars, compiled by Lunar and Planetary Institute, LPI Contribution 441 (1981)

(28) Tobias Owen et al., "The Composition of the Atmosphere at the Surface of Mars," J. Geophys. Res. 82:4635-4639 (1977)

(29) Fraser P. Fanale, "Mars Climate Change: Where are the Petroglyphs?", Nature 294:5839 pp. 308-310 (1981)

(30) D. Nummedal, "Continental margin sedimentation: its relevence to the morphology on Mars," Reports of Planetary Geology Program, NASA Technical Memorandum 85127 (1982)

(31) Carlton C. Allen, "Volcano-ice interactions on Mars," J. Geophys. Res. 84:8048-8060 (1979)

(32) Carroll Ann Hodges and Henry J. Moore, "The subglacial birth of Olympus Mons and its aureoles," J. Geophys. Res. 84:8061-8074 (1979)

(33) P. Toulmin et al., "Geochemical and mineralogical interpretation of the Viking inorganic chemical results," J. Geophys. Res. 82:4625-4634 (1977)

(34) J. L. Gooding and K. Keil, Geophys. Res. Lett., 5, pp. 996-998 (1978)

(35) A. K. Baird et al., J. Geophys. Res. 82:4595-4624 (1977)

(36) Francis J. Pettijohn, Sedimentary Rocks, Third Edition, Harper & Row, p. 591 (1975)

(37) Robert L. Huguenin, "The Formation of Goethite and Hydrated Clay Minerals on Mars," J. Geophys. Res. 79:3895-3905 (1974)

(38) Richard V. Morris and H. V. Lauer, Jr., "The Case against UV photostimulated oxidation of magnetite," Geophys. Res. Lett. 7(8):605-608 (1980)

(39) Carl Sagan et al., "Total reflection spectrophotometry and thermogravimetric analysis of simulated Martian surface materials," Icarus 4:43-61 (1965)

(40) Preston Cloud, Chapter 10, "How the Air became Breathable," pp. 117-140 in Cosmos, Earth and Man, Yale University Press (1978)

(41) Humberto Maturana and Heinz von Foerster (private communication)

(42) Biospherics Inc., "Development of Biological and Non-biological Explanations for the Viking Labeled Release Data," Final Report, 15 Feb., 1979-1980 NASA-CR-163291

(43) Hugh H. Kieffer, "Mars south polar spring and summer temperatures: a residual carbon dioxide frost," J. Geophys. Res. 84:8263-8288 (1979)

(44) J. C. Pechmann, "The origin of polygonal troughs on the northern plains of Mars," Icarus 2:2, pp. 185-210 (1980)

(45) James A. Cutts and Blake H. Lewis, "Models of climate cycles recorded in Martian polar layered deposits," Icarus 50(2/3):216-244 (1982)

(46) Haim Tsoar et al., "Mars: The north polar sand sea and related wind patterns," J. Geophys. Res. 84:8167-8182 (1979)

(47) Carol S. Breed et al., "Morphology and distribution of common 'sand' dunes on Mars: Comparison with the earth," J. Geophys. Res. 84:8183-8204 (1979)

(48) Wesley A. Ward, "Yardangs on Mars: Evidence of recent wind erosion," J. Geophys Res. 84:8147-8166 (1979)

(49) J. F. McCauley et al., "Pitted and fluted rocks in the western desert of Egypt: Viking comparisons," J. Geophys. Res. 84:8205-8221 (1979)

(50) Carl Sagan, Cosmic Connection, Doubleday (1975)

(51) El-Baz et al., "Eolian features in the western desert of Egypt and some applications to Mars," J. Geophys. Res. 84:8295-8221 (1979)

(52) D. C. Pieri, "Geomorphology of Martian valleys," Ph.D. diss., Cornell Univ. (1979)

(53) Michael H. Carr, The Surface of Mars, p. 138, Yale University Press (1981)

(54) K. L. Jones, "Evidence for an episode of Martian crater obliteration intermediate in Martian history," J. Geophys. Res. 79:3917-3932 (1974); C. R. Chapman and K. L. Jones, "Cratering and obliteration history on Mars," Ann. Rev. Earth Planet. Science. 5:515-540 (1977)

(55) L. Alvarez et al., Science 1984

(56) Michael H. Carr (private communication)

(57) Bruce A. Smith, "NASA Studying Possibility of Second Galileo Vehicle," Aviation Week & Space Technology, Jan. 31, 1983

(58) U. Vonzahn and W. Kokott, "Study of Mission Parameters for an ESA Mars Geophysical Orbiter ('Kepler')," Final Report, NASA BMFT-FB-W-028; ISSN-0170-1339 Dec, 1981

(59) Michael H. Carr (private communication)

(60) "Soviet Union to Zap Martian Moon with Laser-equipped Robot," AP wirestory, June 1984

Acknowledgments. The author wishes to thank the following individuals for a variety of contributions and support, without whom this research would have been impossible:

R. Blumenthal, J. Brandenburg, R. Breck, M. Carr, J. Channon, M. Davies, V. DiPietro, D. Drasin, L. Dolphin, M. Evren, B. Graves, H. Knox, R. Patterson, R. Pozos, T. Rautenberg, P. Shay, S. Strand, and D. Webb.

- O -

152

ADDITIONAL READING

Aquinas, Thomas *Summa Theologica.*

Aristotle *Metaphysica.*

Aristotle *Poetica.*

Atlas of Mars, Government Printing Office, Washington, D.C.

Aveni, Anthony F. and Gary Urton, eds. 1982 *Ethnoastronomy and Archaeoastronomy in the American Tropics.* Annals of the New York Academy of Sciences. Volume 385. New York: The New York Academy of Sciences.

Baker, Victor R. 1982 *The Channels of Mars,* Univ. of Texas Press, Austin.

Bateson, Gregory 1971 (1972) ''The Cybernetics of 'Self': A Theory of Alcoholism.'' Reprinted in *Steps to an Ecology of Mind.* New York: Ballantine Books, pp. 309-337.

Bateson, Gregory 1972 ''From Versailles to Cybernetics.'' 1966 lecture first printed in *Steps to an Ecology of Mind.* New York: Ballantine Books, pp. 469-477.

Beatty, J. Kelly, Brian O'Leary and Andrew Chaikin, eds. 1981 (1982) *The New Solar System.* Second edition. Introduction by Carl Sagan. Cambridge, Massachusetts; Cambridge University Press and Sky Publishing Corporation.

Benedict, Ruth 1946 *The Chrysanthemum and the Sword: Patterns of Japanese Culture.* Boston: Houghton Mifflin Company.

Boslough, John 1985 *Stephen Hawking's Universe.* New York: William Morrow and Company, Inc.

Bradbury, Ray 1958 *The Martian Chronicles.* Garden City, N.Y.: Doubleday.

Bronowski, Jacob 1973 (1974) *The Ascent of Man.* Boston: Little, Brown.

Burgess, Eric 1978 *To The Red Planet.* New York: Columbia University Press.

Capra, Fritjof 1975 *The Tao of Physics: An Exploration of the Parallels Between Modern Physics and Eastern Mysticism.* Berkeley: Shambhala. (New York): distributed in U.S. by Random House.

Capra, Fritjof 1982 (1983) *The Turning Point: Science, Society, and the Rising Culture.* New York: Bantam Books.

Carr, Michael H. 1981 *The Surface of Mars.* New Haven: Yale University Press.

Carroll, Lewis 1893 *Alice's Adventures in Wonderland.* New York, Boston: T.Y. Crowell & Co.

Chapman, Clark R. 1977 *The Inner Planets,* Chas. Scribner Sons, N.Y.

Churchman, C. West 1979 *The Systems Approach and Its Enemies.* New York: Basic Books, Inc.

Clarke, Arthur C. 1968 *2001: A Space Odyssey.* London: Hutchinson. — Novel based on screenplay by Arthur C. Clarke and Stanley Kubrick.

Cohn, Norman 1970 (1972) *The Pursuit of the Millenium.* New York: Oxford University Press.

Dante Alighieri 1265-1321 *Divina Commedia.* [*The Divine Comedy.*]

Domhoff, G. William 1967 *Who Rules America?* Englewood Cliffs, N.J.: Prentice-Hall.

Einstein, Albert 1920 *Relativity: The Special and General Theory.* Robert W. Lawson, trans. New York: H. Holt and Company.

Ferguson, Marily 1980 *The Aquarian Conspiracy: Personal and Social Transformation in the 1980s.* Los Angeles, J.P. Tarcher; (New York): distributed by St. Martin's Press.

Geertz, Clifford 1966 (1973) ''Religion As a Cultural System.'' in *The Interpretation of Cultures.* New York: Basic Books, pp. 87-125.

Grossinger, Richard 1981 *The Night Sky.* San Francisco: Sierra Club Books.

Henbest, Nigel 1981 *The Mysterious Universe.* London: Ebury Press.

Journal of Geophysical Research, Vol 82, No. 28, Sept. 30, 1977. ''Scientific Results of the Viking Project.''

Kant, Immanuel 1724-1804 *Critique of Pure Reason.*

Lewis, Clive Staples 1943 *Out of the Silent Planet*. New York: The Macmillan Company.

Levi-Strauss, Claude 1949 (1973) *Tristes Tropiques*. John & Doreen Weightman, trans. London: Cape.

Liebes, Sidney Jr. *The Planets*. Sci. Am. Library, 1975-1983.

Lowell, Percival 1895. *Mars*. Boston & New York: Houghton, Mifflin & Company.

MacDonald, Gordon A. 1972. *Volcanoes*. Prentice-Hall, Englewood, N.J.

May, Rollo 1975 *The Courage to Create*. New York: Norton.

May, Rollo 1969 *Love and Will*. New York: Norton.

May, Rollo 1972 *Power and Innocence: a search for the sources of violence*. New York: Norton.

Mitroff, Ian 1983 *The Subjective Side of Science*. Seaside, CA: Intersystems Publications.

Moore, Patrick, 1977 *Guide to Mars* W.W. Norton Co., N.Y.

Morgan, Lewis Henry 1877 (1964) *Ancient Society*. Cambridge: Belknap Press of Harvard University Press.

Murray, Bruce, ed. 1975 (1977, 1978, 1979, 1980, 1981, 1982, 1983) *The Planets*. Readings from *Scientific American*. San Francisco: W.H. Freeman and Company.

Mutch, Thomas A. et. al., 1976 *The Geology of Mars*.

NASA-CR-164767, *Third International Colloquium on Mars,* Pasadena, Aug 31 - Sept 2, 1981.

Sagan, Carl 1980 *Cosmos*. New York: Random House.

Soleri, Paolo 1969 *Arcology: The City in the Image of Man*. Cambridge, Mass.: The MIT Press.

Spitzer, Cary R., ed. 1980 *Viking Orbiter Views of Mars*. Washington, D.C.: National Aeronautics and Space Administration. (NASA SP-441).

Stanley, Steven M. 1981 *The New Evolutionary Timetable: Fossils, Genes, and the Origin of Species*. New York: Basic Books, Inc.

Tylor, Sir Eward B. 1874 (1958) *Primitive Cultures*. New York: Harper.

Well, W.A. 1979 *Geophysics of Mars*. Elsevier Scientific Publishing Co., N.Y.